TOURISM IN TURKEY

A Comprehensive Overview and Analysis
for Sustainable Alternative Tourism

Advances in Hospitality and Tourism

TOURISM IN TURKEY

A Comprehensive Overview and Analysis for Sustainable Alternative Tourism

Edited by

Ahmet Salih İkiz, PhD

APPLE ACADEMIC PRESS

First edition published 2022

Apple Academic Press Inc.
1265 Goldenrod Circle, NE,
Palm Bay, FL 32905 USA

4164 Lakeshore Road, Burlington,
ON, L7L 1A4 Canada

CRC Press
6000 Broken Sound Parkway NW,
Suite 300, Boca Raton, FL 33487-2742 USA

2 Park Square, Milton Park,
Abingdon, Oxon, OX14 4RN UK

Apple Academic Press exclusively co-publishes with CRC Press, an imprint of Taylor & Francis Group, LLC

Library and Archives Canada Cataloguing in Publication

Title: Tourism in Turkey : a comprehensive overview and analysis for sustainable alternative tourism / edited by Ahmet Salih İkiz.

Names: Ikiz, Ahmet Salih, 1968- editor.

Series: Advances in hospitality and tourism book series.

Description: First edition. | Series statement: Advances in hospitality and tourism | Includes bibliographical references and index.

Identifiers: Canadiana (print) 20200387898 | Canadiana (ebook) 2020038791X | ISBN 9781771889384 (hardcover) | ISBN 9781003084235 (ebook)

Subjects: LCSH: Sustainable tourism—Turkey.

Classification: LCC G156.5.S87 T68 2021 | DDC 338.4/79109561—dc23

Library of Congress Cataloging-in-Publication Data

Names: Ikiz, Ahmet Salih, 1968- editor.

Title: Tourism in Turkey : a comprehensive overview and analysis for sustainable alternative tourism / edited by Ahmet Salih Ikiz.

Description: First edition. | Palm Bay, FL : Apple Academic Press, 2021. | Series: Advances in hospitality and tourism | Includes bibliographical references and index. | Summary: "Tourism in Turkey: A Comprehensive Overview and Analysis for Sustainable Alternative Tourism is an informative study that analyzes the Turkish tourism from a variety of different perspectives and sheds light on its vast economic and social impacts. The volume takes into account Turkey's popularity as a tourism destination that attracts many visitors, compared to the other popular destinations, yet its per capita earnings is low and suffers from declining foreign visitors. This volume emphasizes sustainable alternative tourism as a way forward. The Turkish hospitality and travel sector contributes significantly to the country's GDP and is responsible for a considerable level of employment in Turkey. The volume discusses the development stages of tourism sector in Turkey in the light of policy changes that parallel the economic, political, social, and administrative transformations in Turkey. With the country's favorable location, existing potential, mega projects, and ambitious targets set for 2023, the tourism sector needs to implement sustainable tourism measures in order to grow. The book considers the history of Turkish tourism, regional touristic places and special destinations, government policy and plans, demand pattern of visitors, hospitality and gastronomy sector, coastal marinas and impacts, health tourism, religious tourism, and more. Key features of the volume: Emphasizes the potential of Turkish tourism. Analyzes the main determinations of the tourism market. Explores the importance and branding of Turkish destinations around the globe. Discusses government policies and future expectations in the sector. Considers alternative touristic assets. Provides an overview of Turkish foreign policy in relation to the hospitality sector. This enlightening volume provides an abundance of research that will be valuable for academia and tourism professionals"-- Provided by publisher.

Identifiers: LCCN 2020050394 (print) | LCCN 2020050395 (ebook) | ISBN 9781771889384 (hardcover) | ISBN 9781003084235 (ebook)

Subjects: LCSH: Sustainable tourism--Turkey. | Tourism--Turkey--Planning. | Hospitality industry--Environmental aspects--Turkey.

Classification: LCC G156.5.S87 T857 2021 (print) | LCC G156.5.S87 (ebook) | DDC 338.4/791561--dc23

LC record available at https://lccn.loc.gov/2020050394

LC ebook record available at https://lccn.loc.gov/2020050395

ISBN: 978-1-77188-938-4 (hbk)
ISBN: 978-1-77463-823-1 (pbk)
ISBN: 978-1-00308-423-5 (ebk)

This book is dedicated to Professor Ethem Ruhi Fığlalı who was the Founding Rector of Muğla University 1992–2002

Professor Fığlalı was born on December 8th 1937 in Burdur, Turkey. He was appointed as a Full Professor in Ankara University, Ankara Turkey on August 31st 1982. He became the Vice/Deputy Rector at Dokuz Eylül University in İzmir, Turkey. He also served as the Dean of Faculty (College) of Divinity of the same university until 1991. On November 10, 1992 he was appointed as the Founding Rector of Muğla University, Turkey. He retired on November 10, 2002. He has published many books and articles in the field of Islamic studies and Alevi. He pioneered a major donation to Muğla University from a great philanthropist Sıtkı Davut Koçman. Professor Ethem Ruhi Fığlalı currently lives in Aydın, Turkey.

ABOUT THE EDITOR

Ahmet Salih İkiz, PhD, is on the faculty in the Political Science and International Relations Department at Muğla Sıtkı Koçman University, Turkey. He holds MSc degrees in International Financial Management from Istanbul University and International Trade and Finance from Lancaster University. He completed his PhD on the shadow economy and economic growth in Turkey at Dokuz Eylül University, Turkey. He has published on EU studies, the shadow economy, and the Turkish economy. He had taken research trips to Russia, Kazakhstan, and abroad.

Dr. İkiz can be reached at ahmet@mu.edu.tr.

CONTENTS

CONTRIBUTORS

İlyas Akhisar

Dr. Ilyas Akhisar is working on the Faculty of Economics and Administrative Sciences in Kocaeli University, Turkey. He holds MSc degrees in Mathematical Engineering Department from Istanbul Technical University in Turkey. He completed his PhD on the Mathematical Engineering Department from Istanbul Technical University in Turkey. He has done much research about banking and insurance and also published many books about banking and insurance. E-mail: akhisar@marmara.edu.tr

Umut Avci

Dr. Umut Avci is working as Professor of Hospitality Management in the Department of Business Administration, Faculty of Economics and Administrative Sciences, at Mugla Sitki Kocman University, Turkey. Dr. Avci's research expertise is in hospitality management, strategic orientation, competitive advantage, learning organization, environmental management, and emotional labor. He received his PhD and his MS in Management and Organization on Service Business. He worked in the hotel industry for more than 10 years before becoming an educator. His teaching areas include strategic management, international business, and human resource management. Dr. Avci is the author of several refereed papers in leading academic journals and international conferences. He is on the editorial board of some journals such as *International Journal of Contemporary Hospitality, Journal of Travel and Hospitality Management,* etc. E-mail: aumut@mu.edu.tr

Ozan Bahar

Dr. Ozan Bahar is a Full Professor in the Department of Tourism Faculty at the University of Mugla Sıtkı Kocman, Turkey. Professor Bahar obtained his PhD from the University of Mugla Sıtkı Kocman with a thesis on "An Empirical Investigation of the Competitive Position of the Tourism Sector in Turkey: A Study in Mugla." Professor Bahar's research interests are in the areas of tourism economics, and competitive and sustainable tourism. He sits on the editorial boards of the *Journal of Hospitality and Tourism Insights* and *Journal of Turkish Tourism Research.* He is the coauthor of *Tourism Economics: Concepts and Practices* (2008) and many other books related to the field. He has been a reviewer of many academic journals, including *Tourism Management, Applied Economics, Journal of Business and Economic Management,* etc. He is the first disabled dean of Turkey. E-mail: obahar@mu.edu.tr

Uğur Çalışkan

After working at the Ministry of Culture and Tourism and Regional Development Agencies, Uğur Çalışkan has been working at Muğla Sıtkı Koçman University, Faculty of Tourism (Turkey) as Associate Professor since 2015. He has published many articles in national and international peer review journals, conferences, and book chapters. His research interests include tourist satisfaction, tourist behavior, tourism planning, local residents' reactions toward tourism, entrepreneurship in tourism, tourism sociology, culture and tourism, and commodification. He is currently giving lectures on entrepreneurship in tourism, tourism legislation, management of cultural heritage and tourism, and tourism policy and planning. E-mail: ugurcaliskan@mu.edu.tr

Filiz Gümüş Dönmez

Dr. Filiz Gümüş Dönmez works in the department of Tourism Management, Tourism Faculty, at Muğla Sıtkı Koçman University, Turkey. She completed her PhD and MS in Tourism Management and Bachelor's degree in Tourist Guiding. She had worked 10 years as a tourist guide, restaurant manager, and saleswoman in the tourism industry before working at university. Her research area is in tourism management, workplace loneliness, organizational behavior, sustainable tourism, local residents, and diversification in tourism. She teaches social psychology, behavioral science, organizational behavior, and human resources management. Dr. Gümüş Dönmez has several papers in academic journals, national, and international conferences. E-mail: filizgumus@mu.edu.tr

Gülay Örmeci Güney

Dr. Gülay Örmeci Güney works at Muğla Vocational School (Foreign Trade Department) at Muğla Sıtkı Koçman University, Turkey. She graduated from Istanbul University, Faculty of Economics, Department of Economics. She holds MSc degrees in Graduate School of Social Sciences Department of Economics from Muğla Sıtkı Koçman University. She completed her PhD on "Conspicuous Consumption in Developing Countries: A Case Study in Turkey (Muğla Case)" at Muğla Sıtkı Koçman University, Turkey. She has published on consumption, poverty, and development studies. She had research trips to India, Israel, and Bulgaria. E-mail: hgulay@mu.edu.tr

Nur Çelik İlal

She holds her Bachelor's degree from Mersin University (Turkey), the Department of Tourism Management and Hospitality, and Master's from Mugla Sitki Kocman University (Turkey), the Department of Economics. She has been working as a Research Assistant at Mugla Sitki Kocman University since 2017. Çelik İlal's research interests are in the field of general tourism, tourism economics, and sustainable tourism. E-mail: nur-48@hotmail.com

Asli Karataş

Dr. Aslı Karataş works at Muğla Vocational School (Department of Marketing and Advertising) at Muğla Sıtkı Koçman University, Turkey. She graduated from Muğla Sıtkı Koçman University, Faculty of Economics, Department of Economics. She holds an MSc degree from the Graduate School of Social Sciences Department of Economics from Muğla Sıtkı Koçman University. She completed her PhD on "Gender Career and Wage Gap in the Banking Sector in Istanbul" at Muğla Sıtkı Koçman University, Turkey. She has published on woman labor, sustainable development, economic growth, and development studies. E-mail: aslikaratas@mu.edu.tr

Levent Karadağ

Dr. Levent Karadağ works at Datça Kazım Yılmaz Vocational School at Muğla Sıtkı Koçman University, Turkey. He holds MSc degrees in Tourism Management from Muğla Sıtkı Koçman University. He completed his PhD on sustainable tourism and alternative tourism in Turkey at Muğla Sıtkı Koçman University, Turkey. He has published on sustainable tourism, tourist decision-making process, alternative tourism, and tourism management. E-mail: leventk@mu.edu.tr

Hasan Köşker

Dr. Hasan Köşker works at Kdz. Eregli Tourism Faculty, Zonguldak Bulent Ecevit University, Turkey. He holds MSc degrees in Tourism Management from Istanbul University. He completed his PhD on research on "the basic personality traits of tourism undergraduate students and their service orientation with their traits towards working tendency in tourism sector" Mersin University, Turkey. He has published on geographical indication, service orientation, mountain tourism, cultural tourism, and Turkish tourism economy. E-mail: hasankosker@beun.edu.tr; hasankosker@yahoo.com

Zeynep Kurtulay

Zeynep Kurtulay works as Research Assistant in the Kdz. Eregli Tourism Faculty, Zonguldak Bülent Ecevit University, Turkey. She holds an MSc degree in Tourism Management from İstanbul University and is a PhD student in Tourism Management at İstanbul University, Turkey. She works on fuzzy logic, marketing, and AI. E-mail: z.kurtulay@beun.edu.tr

Merve Tuna Mutlu

Merve Tuna Mutlu is a PhD student at the English Teaching Department, Middle East Technical University and research assistant at the Muğla Sıtkı Koçman University. She has an MA degree in English Teaching from Anatolian University. She already passed preliminary exam and preparing dissertation proposal. Her main area of interest is alternative English teaching strategies for older age population. E-mail: muammert@hotmail.com

Nilüfer Vatansever Toylan

Dr. Nilüfer Vatansever Toylan works as Tourism Faculty member in Tourism Management Department at Kırklareli University, Turkey. She holds an MSc degree in Business Administration from Hacettepe University and completed her PhD on the strategic alliances, knowledge sharing, and innovation in hotels in Turkey at İstanbul University, Turkey. Dr. Toylan continues to research in the subjects of tourism management, strategic management, contemporary management, interorganizational and organizational trust, knowledge management, and knowledge sharing and innovation. E-mail: nilufer.vatansever@klu.edu.tr

Muammer Tuna

Muammer Tuna, PhD, is a Professor of Sociology. His main interest areas are environment, tourism, modernization and development, migration, and sociology of aging. He has done a BA in sociology from Ankara University, MA in Turkish History from Yıldız Technical University, and PhD in sociology from Mississippi State University. He has published on environmental sociology, sociology of tourism and environment, migration, and aging and directed and participated number of researches on mainly environment, tourism, and migration. He is currently Chair of Department of Sociology at the Muğla Sıtkı Koçman University, Muğla, Turkey. E-mail: muammert@hotmail.com

Uygar Turgut

Uygar Turgut is an English teacher at Koycegiz District Directorate of National Education as a project coordinator. During his 15-year career teaching at state schools, he has coordinated different European School Projects such as Comenius and Erasmus+. He completed a Masters of Arts degree in Tourism Management at Mugla Sıtkı Kocman University. He is now doing his doctorate in tourism management at the same university. E-mail: uygarturgut@yahoo.co.uk

ABBREVIATIONS

DSP	dominant social paradigm
GDP	gross domestic product
GNP	gross national product
IHEI	International Hotels Environment Initiatives
ITA	Istanbul Travel Agency
LDA	linear discriminant analysis
Natta	National Turkish Tourist Agency
MANOVA	multivariate analysis of variance
MDA	multiple discriminant analysis
SEM	structural equation modeling
SPO	State Planning Organization
TTCI	travel and tourism competitiveness index
T&T	tourism and travel
TST	Tourism Strategy of Turkey
Tutta	Turkish Traveling & Tourist Agency
UNWTO	United Nations World Tourism Organization
WTO	World Trade Organization
WTTC	World Travel and Tourism Council
WWOF	Worldwide Opportunities on Organic Farms

PREFACE

The tourism industry has global dimensions due to its international demand patterns. Thus, we have compiled a book with an international well-known publisher. Researchers all around the world working on Turkish tourism industry can use it.

The Turkish tourism industry has grown after 1980s with the liberalization of economic policies with increasing efforts to export to the global market. Increasing tourism activity, in turn, attracted academia on these studies. Those publications were mainly textbooks for undergraduate students in universities. In last couple of decades, the increasing internationalization of academia has increased the number of publications in foreign languages. So both undergraduate and graduate students can benefit from this book, both in Turkey and abroad.

Even if there is an increasing trend in publication on the tourism industry in Turkey, there are still not too many international books on this local issue around the globe. Therefore, international academicians who are doing scientific research on the Turkish tourism industry in universities and in different parts of the world may take advantage of this book. We wish all readers to enjoy the updated knowledge provided in this book.

INTRODUCTION

In today's global village, leisure activities are important aspects of daily life. As with most of the developing world, Turkey has an important amount of gains from the tourism industry. At the macroeconomic level, it performs like a smokeless factory that exports service products all around the world. It contributes to both GDP and employment in a country like Turkey where it manages to cure current account deficits of balance of payment while diminishing the unemployment level.

Although Turkey is a very popular tourist destination and attracts many visitors, compared to the other popular destinations in the globe, the per capita earnings are low and suffer from declining foreign visitors. Therefore, income generation and multiplier effect of tourism revenues do not increase incrementally with higher number of foreign visitors to the country. The low segment tourist profile curtails the potential benefits of the industry.

One of the main arguments of the tourism industry is about its sustainability. Since the main assets of the tourism industry rely on natural resources and cultural heritage, those assets can be exhausted with overutilization. Thus, sustainable alternative tourism became a crucial part of the new tourism trend in order to both prevent adverse effects of mass tourism and for the creation of new markets by alternative tourism.

In this study, we have attempted to draw the situation analyses of the tourism industry in a comprehensive manner of Turkey. How we can improve the potential gains from the sector and by which means were main questions. Thus, we put some policy recommendations and remedies to the current situation of the Turkish tourism industry.

CHAPTER 1

GENERAL FRAMEWORK OF TURKISH TOURISM: TOURISM AND TRAVEL COMPETITIVENESS INDEX AND TURKEY

A. S. İKİZ

Faculty of Economics & Administrative Studies, Muğla Sıtkı Koçman University, Turkey

**Corresponding author. E-mail: ahmet@mu.edu.tr*

ABSRACT

Turkish tourism industry plays crucial role in macroeconomic context in country. Thus, it became a main remedy for acute current account deficits. But the number of incoming tourists cannot reflect the revenues of sector. Even though there is sharp increase in the number of visitors, the average spending is declining. In this paper, we fist define travel competitiveness index and effects of Turkish tourism sector. We then try to evaluate Turkish tourism industry output via this index.

1.1 INTRODUCTION

Travelers and tourists spend their money on a vast variety of goods and services in host countries. These expenditures are main revenues for the host country by creating income injection under economic bases. These revenues depend upon income creation by foreign currencies brought to the country for touristic purposes. In this chapter, we first provide information about the general framework of the Turkish tourism industry.

Global competition is a vital part of the industry, so the travel and tourism competitiveness index (TTCI) plays a vital role in evaluating countries' potential from an industry point of view. In the following part, the TTCI pillars and the Turkish tourism industry's strengths and weaknesses are briefly explained. In the last part of the chapter, policy recommendations for the Turkish tourism industry are discussed.

1.2 ECONOMIC GROWTH AND TOURISM REVENUES IN TURKEY

Tourism is classified as the second largest industry in the global context. It creates jobs, employment more than most industrial production in some cases. The direct and indirect impacts of income generation by tourism expenditures may easily be traced in the following chart.

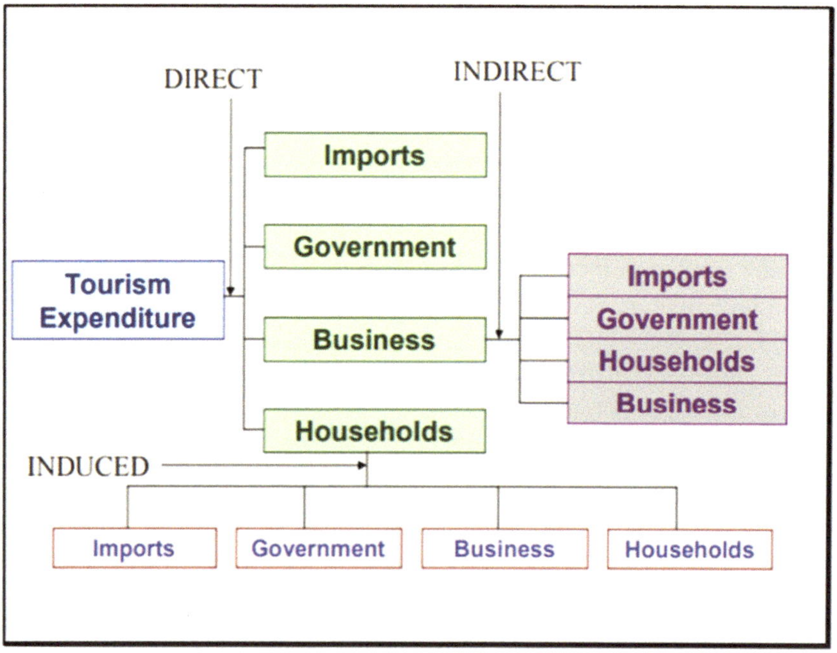

FIGURE 1.1 The effects of tourism expenditure.

Source: Ennew (2003),http://citeseerx.ist.psu.edu/viewdoc/download?doi=10.1.1.567.8713&rep=rep1&type=pdf.

Tourism is one of the fastest-growing sectors in the world in recent years. Its linkages with other sectors are quite strong in the gross domestic product (GDP) generation process. Tourism-led growth thus has important aspects of the globe.

Some studies have found no evidence of tourism-led growth in Turkey (Öztürk and Acaravcı, 2009). One may easily infer that the low level of the spending pattern of tourists creates such a pattern. Tourism revenues make a positive contribution to economic growth in most developing countries (Ekanayake and Long, 2012). Tourism revenues have positive impacts on the balance of payments in the service sector part and create surplus in foreign reserves that most developing countries have. The increase in these revenues may accelerate economic growth in these countries as well.

Thus, governments' policy choice on attracting premier selective tourists in developing countries would promote tourism as a potential source of economic growth. In the meantime, most developing countries need financial capital to fund their development. Tourism seems to be a good option to increase the amount of foreign currency needed to deal with this problem. For this reason, it appears that developing nations are focused on policies to increase the number of foreign guests in their countries (Dincer et al., 2015).

The Turkish tourism industry has had much more vital importance in the GDP after 1980 by increasing private investments in the sector (Keskin and Cansız, 2010). Since foreign tourists' spending generates more direct and induced income on GDP, it has important aspects for Turkish economy (Liu et al., 1984).

Tourism-led growth hypothesis for Turkish economy reveals that both international tourist arrivals and international tourist expenditures positively affect the economic growth in Turkey. The results indicate that, in the long run, economic growth in Turkey is strongly influenced from the tourism expansion policies of the respective governments (Savaş et al., 2010). The multiplier effect of tourism spending is quite strong compared to many countries. It also generates at least 5% of the Turkish GDP with foreign earnings (Uşaklı and Ünlüönen, 2015). Thus, it is one of the main remedies of the acute current account deficits of Turkish economy for years.

Since Turkey is a labor-abundant country, the labor-intensive tourism industry creates employment with zero opportunity cost (Akal, 2010). The increase in tourism revenues has a significant impact on economic growth in Turkey (Terzi, 2015). However, we have to bear in mind that economic growth with a positive current account balance has a vulnerable structure that is affected dramatically by entire crises and comprises sudden changes and conversions within (Alp and Genc, 2015). Some research findings advocate that immiserating growth implications of the tourism industry are quite high in so many countries (Sarıdoğan and Çetin, 2016). Tourism revenues increase economic growth, but also vice versa, where economic growth may cause positive impacts on tourism revenues (Samırkaş and Samırkaş, 2014).

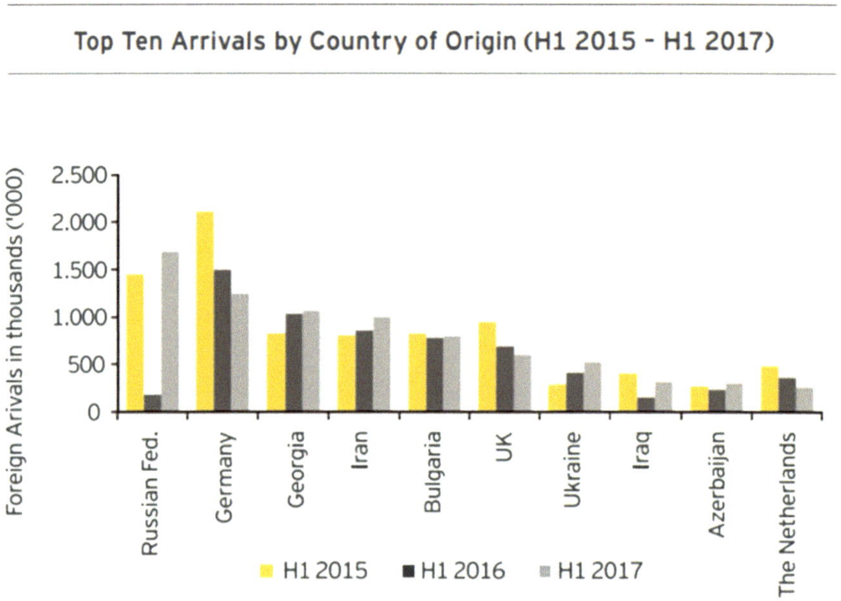

FIGURE 1.2 Arrivals to Turkey from country of origin.

Source: EY Report (2017), https://www.ey.com/Publication/vwLUAssets/Tourism/%24FILE/EY_Tourism_Industry_Overview_2017.pdf.

The European and Western world dominated the profile of foreign visitors in the past millennium, whereas, in the post-2002 period, there has been a changing trend in tourists in favor of the Russian Federation. In the last decade, there has been a sharp increase in the share of visitors from the Arab world and the Middle East.

After a crisis in the year 2015, the number of foreign visitors soared with 21.84% by 39.5 million in 2018 compared to the previous year. Last year, nearly 6 million Russians visited Turkey, by a 26.5% increase from 2017, while the number of German tourists rose by 25.9% on an annual basis to 4.5 million. The third-largest group was Bulgarians. Visits from this country were up 28.8% and 2.4 million. In 2018, 2.25 million Britons spent their holidays in Turkey, by a 35.9% increase from 2017 (Hurriyet Daily News, 2019). The per capita spending of foreign visitors to Turkey reached $600 in 2011 from $250 in 1980 (Akın et al., 2012). However, we have to bear in mind that, in global standards, these figures reveal a low-segment demand. In the meantime, there is a decline in the spending patterns of foreigners. The average expenditure per capita was $647 in 2018, as foreigners spent $617 per capita, and Turkish citizens spent $801 per capita.

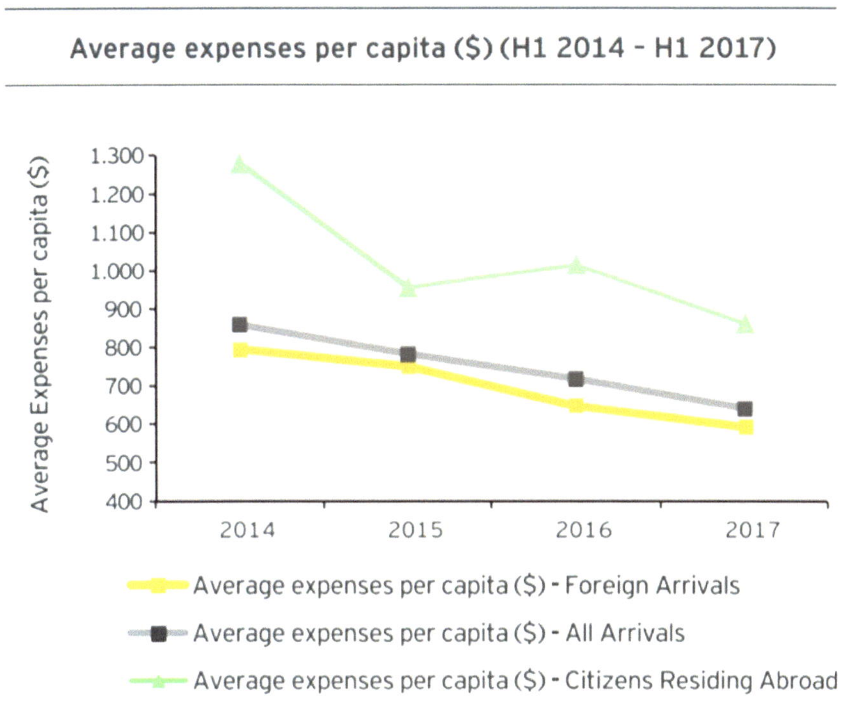

FIGURE 1.3 Average per capita expenses of visitors to Turkey.

Source: EY Report (2017), https://www.ey.com/Publication/vwLUAssets/Tourism/%24FILE/EY_Tourism_Industry_Overview_2017.pdf.

The global tourism trends are flourishing with the increasing number of travelers and increasing earnings from this industry, whereas these global trends have different outcomes in Turkey with increasing number tourists

TABLE 1.1 Tourism Receipts and Expenses of Turkey.

Years	Tourism receipts and expenditures					
	Inbound visitors	**Outbound visitors**	**Tourism revenues ($1000)**	**Per capita tourism receipt ($)**	**Tourism revenues/ GDP (%)**	**Outbound tourism expenses (million $)**
2003	16,463,623	16,302,053	13,854,866	850	4.4	2425
2004	20,753,734	20,262,640	17,076,607	843	4.2	2954
2005	25,045,142	24,124,501	20,322,111	842	4.1	3395
2006	23,924,023	23,148,669	18,593,951	803	3.4	3271
2007	27,239,630	27,214,988	20,942,500	770	3.1	4043
2008	31,137,774	30,979,979	25,415,067	820	3.3	4266
2009	31,759,816	32,006,149	25,064,482	783	3.9	5090
2010	32,997,308	33,027,943	24,930,997	755	3.2	5875
2011	36,769,039	36,151,328	28,115,692	778	3.4	5531
2012	37,715,225	36,463,921	29,007,003	795	3.3	4 593
2013	39,860,771	39,226,226	32,308,991	824	3.4	5254
2014	41,627,246	41,415,070	34,305,903	828	3.7	5470
2015	41,114,069	41,617,530	31,464,777	756	3.7	5698
2016	30,906,680	31,365,330	22,107,440	705	2.6	5050
2017	37,969,824	38,620,346	26,283,656	681	3.1	5137
2018	46,112,592	45,628,673	29,512,926	647	3.8	4896
2019	41,564,536	41,018,702	26,634,764	649	-	3380

Source: Ministry of Culture and Tourism (2019), https://www.ktb.gov.tr/EN-248841/tourism-receipts-and-expenditures.html.

and declining per capita earnings from inbound travelers. Even though Turkish economy is generating its revenues from tourism, economic boost is accompanied by the declining expenditures from the low spending patterns of travelers.

According to the World Tourism Organization statistics, Turkey received the sixth highest number of arrivals in the world in 2018. However, this figure did not have positive reflections on tourism earnings, since Turkey is not among the top 10 tourism earners in the globe (UNWTO, 2019).

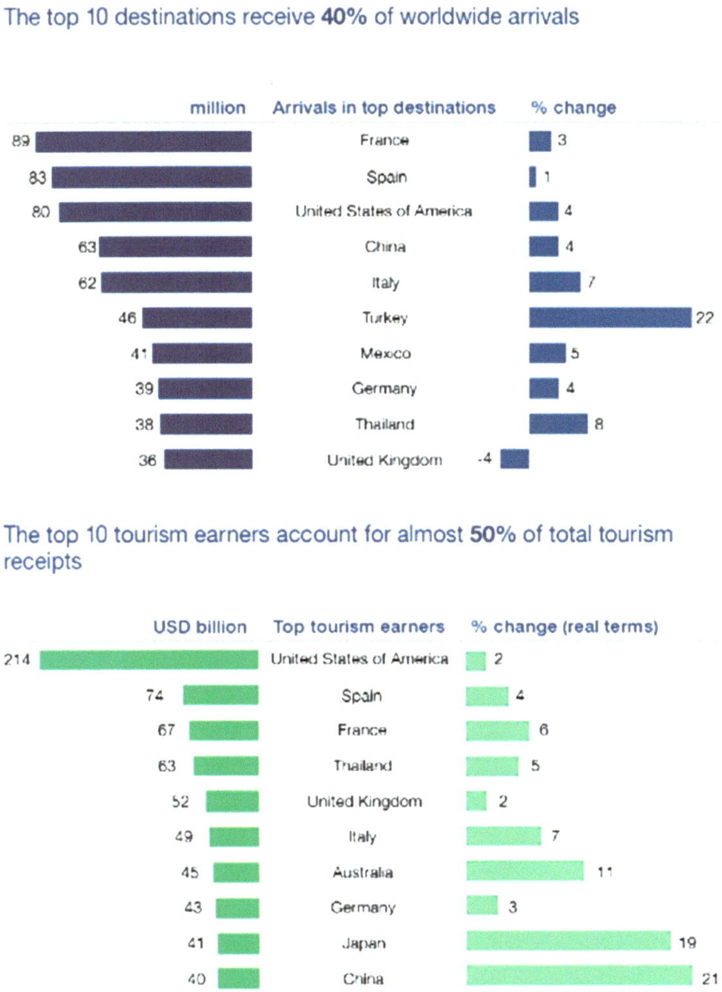

FIGURE 1.4 Top ten destinations and tourism receipts.

The country image highly affects tourism earnings. The image is constructed by both cognitive and affective evaluation, where cognitive evaluation is based on beliefs and knowledge, and affective evaluation is based on feelings (Ayyıldız and Turna, 2014). Tourism demand depends on emotional values with affective evaluation. So, mass media and other global networks highly influence the Turkish image in the globe. The outreach to high segments in tourism demand is clearly related to this issue.

1.3 TOURISM AND TRAVEL COMPETITIVENESS INDEX AND TURKEY

The strength of countries' tourism industries is evaluated by the TTCI produced by the World Economic Forum. The Travel and Tourism Competitiveness Report is a benchmarking tool of cross-country comparison for investors. It helps local governments for policy formation for the promotion of the tourism sector. TTCI comprises 4 subindices, 14 pillars, and 90 individual indicators, distributed among the different pillars. The subindices are as follows (World Economic Forum, 2019):

1. **Enabling Environment** subindex captures the general conditions necessary for operating in a country and includes five pillars:
    a. **Business Environment:** This pillar captures the extent to which a country has in place a policy environment conducive for companies to do business.
    b. **Safety and Security:** Safety and security are critical factors determining the competitiveness of a country's tourism and travel (T&T) industry since tourists are likely to be deterred from traveling to dangerous countries or regions, making it less attractive.
    c. **Health and Hygiene:** Factors such as access to improved drinking water and sanitation are important for the comfort and health of travelers.
    d. **Human Resources and Labor Market:** High-quality human resources in an economy ensure the better quality of service, which is an essential part of this industry.
    e. **ICT Readiness:** Information and communication technology capacities of countries play a vital role in the tourism and travel industry.

2. **T&T Policy and Enabling Conditions** subindex captures specific policies or strategic aspects that affect the T&T industry more directly and includes four pillars:
    a. **Prioritization of Travel and Tourism:** Government policies and efforts for the tourism industry have significant impacts on best practices.
    b. **International Openness:** Developing a competitive T&T sector internationally requires a certain degree of openness and travel facilitation with liberal economic policies.
    c. **Price Competitiveness:** Lower costs related to travel in a country increase its attractiveness for many travelers as well as for investing in the T&T sector
    d. **Environmental Sustainability:** The importance of the natural environment for providing an attractive location for tourism cannot be overstated, so policies and factors enhancing environmental sustainability are an important competitive advantage in ensuring a country's future attractiveness as a destination.
3. **The Infrastructure** subindex captures the availability and quality of physical infrastructure of each economy and includes three pillars:
    a. **Air Transport Infrastructure:** Air connectivity is essential for travelers' ease of access to and from countries, as well as movement within many countries.
    b. **Ground and Port Infrastructure:** The availability of efficient and accessible transportation to key business centers and tourist attractions is vital for the T&T sector. This requires a sufficiently extensive road and railroad network, proxied by road and railroad densities, as well as roads, railroads, and ports infrastructure that meets international standards of comfort, security, and modal efficiency.
    c. **Tourist Service Infrastructure:** The availability of sufficient quality of accommodation, resorts, and entertainment facilities can represent a significant competitive advantage for a country.
4. **The Natural and Cultural Resources** subindex captures the principal "reasons to travel" and includes two pillars:
    a. **Natural Resources:** Countries with natural assets such as UNESCO natural World Heritage sites or richness of the fauna and flora clearly have a competitive advantage in attracting tourists.

TABLE 1.2 Travel and Tourism Competitiveness Index.

	Country/economy	Value (1-7) best
1	Spain	**5.4**
2	France	**5.4**
3	Germany	**5.4**
4	Japan	**5.4**
5	United States	**5.3**
6	United Kingdom	**5.2**
7	Australia	**5.1**
8	Italy	**5.1**
9	Canada	**5.1**
10	Switzerland	**5.0**
11	Austria	**5.0**
12	Portugal	**4.9**
13	China	**4.9**
14	Hong Kong SAR	**4.8**
15	Netherlands	**4.8**
16	Korea, Rep.	**4.8**
17	Singapore	**4.8**
18	New Zealand	**4.7**
19	Mexico	**4.7**
20	Norway	**4.6**
21	Denmark	**4.6**
22	Sweden	**4.6**
23	Luxembourg	**4.6**
24	Belgium	**4.5**
25	Greece	**4.5**
26	Ireland	**4.5**
27	Croatia	**4.5**
28	Finland	**4.5**
29	Malaysia	**4.5**
30	Iceland	**4.5**
31	Thailand	**4.5**
32		
33	UAE	**4.4**
34	India	**4.4**
35	Malta	**4.4**
36	Slovenia	**4.3**
37	Taiwan, China	**4.3**
38	Czech Republic	**4.3**
39	Russian Federation	**4.3**
40	Indonesia	**4.3**
41	Costa Rica	**4.3**
42	Poland	**4.2**
43	Turkey	**4.2**
44	Cyprus	**4.2**

b. **Cultural Resources and Business Travel:** A country's cultural resources are another critical driver of T&T competitiveness such as the number of large stadiums that can host significant sport or entertainment events, and a new measure of digital demand for cultural and entertainment—the number of online searches related to a country's cultural resources can allow the level of interest to be inferred.

In fact, Turkey is far behind many countries with its scores. This simply reveals low per capita earnings from foreign visitors.

Even if Turkey receives millions of tourists from around the world, its TTCI is not quite strong. Its strength is based on price competitiveness, health and hygiene, prioritization of tourism, and infrastructure. On the other hand, the country has low index figures in natural resources, environmental sustainability, ground and port structure, and international openness.

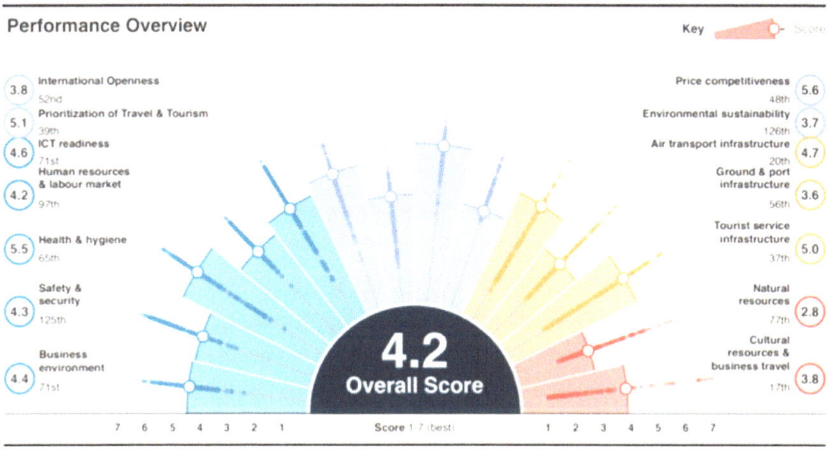

Past performance

Travel & Tourism Competitiveness Edition	2015	2017	2019
Rank	44 / 141	44 / 136	43 / 140
Score	4.1	4.1	4.2

FIGURE 1.5 Turkey TTCI performance.

Source: World Economic Forum (2019), http://reports.weforum.org/travel-and-tourism-competitiveness-report-2019/country-profiles/#economy=TUR.

1.4 POLICY SETS FOR IMPROVING TOURISM REVENUES IN TURKEY

Tourism trends change because of increasing socioeconomic levels, technological developments, and psychological factors affecting tourist behaviors, moderate policies, and new visa procedures, and so on. It is almost impossible that all tourism trends will be popular in every country because of cultural, geographic, social, political, economic, and religions differences (Tanrısever et al., 2016).

Economic factors, political stability, and terrorism are key factors affecting the international tourism demand (Aktas, 2019). In the case of Turkey, the post-2015 period may be characterized by political factors due to the Russian jet crisis and failed coup attempt. Tourism is highly vulnerable to international policies of ruling governments. Although international conflicts in Turkey highly affect tourist arrival in Turkey, the financial crisis in the country has no significant impact on tourism demand. The ideological shift in Turkey changed the pattern of travelers from the world in favor of the Muslim world (Cankurtaran and Çetin, 2016). In the economic aspect, Turkey has many rivals in the region and cannot attract high-segment tourists to the country. Thus, the demand for Turkey becomes sensitive with increasing costs, which means low-income families would be reluctant to visit in case of high travel expenses. Improvement of service quality would have positive impacts on tourism revenues. Some studies suggest that the policy of low prices may not be effective in tourism marketing for Turkey as there are so many rivals in the market (Abedtalas and Toprak, 2015).

Market segmentation is a strategy any entity in the tourism industry can use to strengthen their competitive advantage by selecting the most suitable subgroup of tourists to specialize in and target (Dolnicar, 2008). The bottlenecks in the Turkish tourism sector are highly related to the demand pattern of low-income household travelers. Therefore, low segment demand is pushing the quality of tourism products in Turkey.

There are a couple of factors effecting low segment demand for Turkey in the tourism industry. First, the share of upper-class visitors to the country is quite low, which in turn reflects a low level of spending. Poor service standards, insufficient infrastructure, and price-oriented competition policy are the main drivers behind this outcome.

The brand image is one of the main components of a country's demand. Thus, we have to improve the brand image of Turkey to attain a high level of visitors. A strong brand image of the country with special features would attract high-segment upper-middle-class tourists to the country. The growing popularity of Turkish soap operas and TV serials in the Mediterranean basin, Balkans, and the Middle East increase Turkey's positive image abroad and have helped boost the number of tourists coming to Turkey.

For many decades, Turkish tourism assets have been based on 3S (Sun, Sea, Sand) marketing. The spending patterns of those visitors are quite low compared to culture-based and alternative tourism assets such as medical tourism. Therefore, there is a high need to reach alternative markets.

Turkish hospitality services are dominantly all-inclusive packages, which in turn diminish visitors spending out of the hotels. Low price competition with all-inclusive travel packages has adverse effects on total tourism revenues. In the meantime, Turkey has much more luxury five-star decorated hotels with very poor service quality compared to its rivals in the Mediterranean basin. Inefficiencies in service quality, in fact, had impediments in marketing strategies, which are concentrated on low-price policies.

1.5 CONCLUSION

The tourism industry is becoming a global phenomenon due to its natural demand patterns. Fierce competition in global level pushes countries to fascinate people all around the world in alternative tourism assets. Turkey is facing this competition with declining per capita spending from foreign tourists to the country. Even though there is an increasing tendency in the number of foreign tourists to Turkey, the tourism revenues are not following the same trend. Low segment demand and other factors has adverse effects on tourism revenues of Turkey.

TTCI is a leading tool for measuring countries rankings in the global tourism industry. The index has pillars from hygiene to price competitiveness with score 1–7. The higher the score the higher the ranking in the globe. Countries with high ranking draw high segment demand tourists, which in turn surge tourism revenues. Turkey ranks 43 in 140 countries with considerably low rankings and its strength depend on price competition. Therefore, the economic welfares of tourism revenues are vulnerable with the price level of tourism assets rather than natural and cultural values of Turkey. Thus, the industry has to create innovative methods and alternative assets in the tourism industry in order to break the vicious circle of low per capita earnings from foreign tourists.

KEYWORDS

- **Turkish tourism industry**
- **travel and tourism competitiveness index**
- **Turkish tourism revenues**

REFERENCES

Abedtalas, M.; Toprak, L. The Determinants of Tourism Demand in Turkey. *J. Econ. Behav. Stud.* **2015,** *7*, 90–105.

Akal, M. Economic Implications of International Tourism on Turkish Economy. *Tourismos* **2010,** 112–152.

Akın, A.; Şimşek, M. Y.; Akın, A. The Importance and Place of Tourism in Economy. *J. Acad. Res. Stud.* **2012,** *4*, 63–81.

Aktas, M. C. *The Effects of Terrorism and Political Instability on International Tourist Arrivals: A Focus on Turkey.* North Carolina University, 2019. https://repository.lib.ncsu.edu/bitstream/handle/1840.20/36513/etd.pdf?sequence=1 adresinden alındı.

Alp, E. A.; Genc, E. G. The Relation between Current Account Deficit and Tourism: The Case of Turkey. *Int. J. Econ. Financ. Issues* **2015,** 448–453.

Ayyıldız, H.; Turna, G. B. Turkey's Image as a Tourist Destination in Northern Europe. *Int. J. Adv. Manage. Econ. Entrep.* **2014,** 1–8.

Cankurtaran, B. S.; Çetin, G. Ideology, Foreign Policy and Tourism: The Case of Turkey-Middle East Relations. *Int. J. Turcol.* **2016,** 45–65.

Dincer, M. Z.; Dincer, F. I.; Ustaoglu, M. Reel Effective Exchange Rate Volatilities Impact on Tourism Sector in Turkey: An Empirical Analysis of 2003–2014. *Procedia Econ. Financ.* **2015,** 1000–1008.

Dolnicar, S. *Market Segmentation in Tourism.* Researchgate, 2008. https://www.researchgate.net/publication/30387969_Market_Segmentation_in_Tourism adresinden alındı.

Ekanayake, E. M.; Long, A. E. Tourism Development and Economic Growth in Developing Countries. *Int. J. Bus. Financ. Res.* **2012,** 51–63.

Ennew, C. *Understanding the Economic Impact of Tourism*, 2003. http://citeseerx.ist.psu.edu/viewdoc/download?doi=10.1.1.567.8713&rep=rep1&type=pdf adresinden alındı.

EY Report. *Tourism Market Overview for Turkey & Istanbul.* EY.com, 2017. https://www.ey.com/Publication/vwLUAssets/Tourism/%24FILE/EY_Tourism_Industry_Overview_2017.pdf adresinden alındı.

Hurriyet Daily News. Foreign Arrivals, Tourism Revenues in Turkey Rose Last Year. *Hürriyet Newspaper*, 2019, January 31. http://www.hurriyetdailynews.com/foreign-arrivals-tourism-revenues-in-turkey-rose-last-year-140903 adresinden alındı.

Keskin, A.; Cansız, H. Tourism Turkey and Economic Development. *Atatürk Üniv. İktisadi İdari Bilimler Dergisi* **2010,** 23–46.
Liu, J.; TurgutVar; AlpTimur. Tourist-Income Multipliers for Turkey. *Tour. Manage.* **1984,** 280–287.
Ministry of Culture and Tourism. *Tourism Statistics.* Ministry of Culture and Tourism, 2019. https://www.ktb.gov.tr/EN-248841/tourism-receipts-and-expenditures.html adresinden alındı.
Öztürk, İ.; Acaravcı, A. On the Causality Between Tourism Growth and Economic Growth: Empirical Evidence from Turkey. *Transylv. Rev. Adm. Sci.* **2009,** 73–81.
Samırkaş, M.; Samırkaş, M. C. The Effect of Tourism Industry on Economic Growth: The Case of Turkey. *J. Fac. Bus. Adm. (İşletme Fakültesi Dergisi)* **2014,** 63–76.
Sarıdoğan, Ö.; Çetin, D. Tourism and Immiserizing Growth. *J. Hum. Soc. Sci. Res.* **2016,** 922–932.
Savaş, B.; Beşkaya, A.; Şamiloğlu, F. Analyzing the Impact of International Tourism on Economic Growth in Turkey. *ZKU J. Soc. Sci.* **2010,** 121–136.
Tanrısever, C.; Pamukçu, H.; Batman, O. New Tourism Trends in the World and Their Adaptations to Turkey. *Gümüşhane Univ. Electron. J. Inst. Soc. Sci.* **2016,** 55–72.
Terzi, H. Is the Tourism Led Growth Hypothesis (TLGH) Valid for Turkey. *J. Dogus Univ.* **2015,** 165–178.
UNWTO. *International Tourism Highlights.* World Tourism Organisation, 2019. https://www.e-unwto.org/doi/pdf/10.18111/9789284421152 adresinden alındı.
Uşaklı, A.; Ünlüönen, K. Turizm gelirlerinin hesaplanmasına yönelik yaklaşımlar. *16. Ulusal Turizm Kongresi Bildiriler Kitabı* (s. 1582–1597). Detay Publications: Ankara, 2015.
World Economic Forum. *Travel & Tourism Competitiveness Index.* World Economic Forum, 2019. http://reports.weforum.org/travel-and-tourism-competitiveness-report-2019/country-profiles/#economy=TUR adresinden alındı.

CHAPTER 2

THE PLACE AND IMPORTANCE OF THE TOURISM SECTOR IN TURKISH ECONOMY

O. BAHAR[1] and N. ÇELIK İLAL[2*]

[1]*Faculty of Economics and Administrative Sciences, Mugla Sıtkı Koçman University, Turkey*

[2]*Faculty of Tourism, Mugla Sıtkı Koçman University, Turkey*

[*]*Corresponding author. E-mail: nurcelik@mu.edu.tr*

ABSTRACT

The tourism industry has a positive effect on balance of payments, improves the transfer of technology and knowledge, encourages foreign investments, acts as a locomotive for the development of countries, and gathers people and societies thus contributing to peace. Tourism is one of the fastest growing sectors of the world economy. According to estimates of the World Travel and Travel Council (WTTC), the travel and tourism (T&T) industry account for 10.4% of total GDP (all economic activity including direct and indirect impacts) and its direct impact will constitute 3.8% of total GDP (WTTC, 2018). It is also estimated that the contribution of T&T in total employment is expected to rise from 9.9% of current total employment to 11.6% of total employment by 2028. All these estimates show that tourism has a high share in output, value added, employment, export and hence tax contribution in an economy. The general objective of this study is to analyze the place of tourism in the Turkish economy and the development of tourism within economic indicators by using secondary data. The current data showing the number of tourists coming to the country, income earned from tourists, amount of expenditure per tourists and the share of tourism from GDP, investments and exports, indicate that tourism is one of the leading sectors in the Turkish Economy.

2.1 INTRODUCTION

The tourism industry has a positive effect on the balance of payments, improves the transfer of technology and knowledge, encourages foreign investments, acts as a locomotive for the development of countries, and gathers people and societies, thus contributing to peace. Tourism is one of the fastest-growing sectors of the world's economy. According to the estimates of the World Travel and Tourism Council (WTTC), the travel and tourism (T&T) industry accounts for 10.4% of the total gross domestic product (GDP) (all economic activity including direct and indirect impacts), and its direct impact will constitute 3.8% of the total GDP (WTTC, 2018). It is also estimated that the contribution of T&T in total employment is expected to rise from 9.9% of the current total employment to 11.6% of the total employment by 2028. All these estimates show that tourism has a high share in output, added value, employment, export, and hence, tax contribution in an economy.

Tourism is regarded as a leading industry for the 21st century due to its significant contribution to countries' economies (Giles and Perry, 1998). The tourism sector, which is one of the most significant sectors and driving power of social, economic, and cultural development, has a significant role, especially for developing countries. During the past two decades, tourism has grown rapidly, and the importance of this sector within the world's economy continues to increase.

The sector continues to grow even in times of crisis and it is seen as an important tool in addressing the bottlenecks and macroeconomic problems that countries face, reducing unemployment, increasing public revenues, addressing resource shortages, stimulating investments and alleviating regional development, and external debt problems. Thus, countries are competing with each other to develop their tourism potential, diversify their products, and obtain the desired tourism income from this sector, which continues to grow by the day (Bahar and Kozak, 2012).

According to the United Nations World Tourism Organization (UNWTO) data, international tourism income reached $1.340 billion, and the number of international tourists reached 1326 million in 2017 (UNWTO, 2018).

Moreover, tourism directly and indirectly creates employment for one in every ten people, makes up 7% of the world's total exports and accounts for a 10th of the GDP produced by the world's countries (UNWTO, 2018). According to the data of the Ministry of Culture and Tourism, in 2018, the number of tourists who visited Turkey was a total of 46,112,592, producing a tourism revenue of $29,512,926.

The Tourism Ministry of Turkey released an official document in 2007, which incorporated an ambitious and extremely detailed plan to achieve 50 million tourist arrivals and revenues of $50 billion by 2023. The objective was stated as follows: "Become one of the top five most visited countries in the world, therefore generating mass tourism revenues." The targets and objectives have already been confirmed by the government in 2018 (The Tourism Ministry of Turkey, 2007).

The purpose of this conceptual study is to examine the place of tourism in Turkish Economy and the development of tourism, by analyzing economic indicators used as secondary data. In the first section, a brief assessment of the basic economic impacts of tourism are undertaken; in the next section, the current economic structure of the Turkish tourism sector is evaluated, and in the final section, some recommendations are presented.

2.2 BASIC ECONOMIC IMPACTS OF TOURISM

The importance of tourism in economic and regional development has grown significantly in recent history. Being one of the driving forces of social, economic, and cultural development, the tourism sector is of great importance in terms of developing countries, in particular. Several studies (Çımat and Bahar, 2003; Bahar and Kozak, 2008; Gökovalı, 2010; Ünlüönen et al., 2015) have reported that tourism has significant economic benefits such as the following:

i) contribution to gross national product (GNP) and balance of payments,
ii) creation of employment opportunities,
iii) generation of additional government revenue,
iv) contribution to economic diversity, and
v) promotion of economic development through various multiplier effects.

Another advantage of the tourism sector is that it provides fast and abundant low-cost employment. Within the scope of an index created with the total investment value of the project and the targeted number of jobs created by the investment, tourism has an index value of 100, while this value is 126.8 in the mining sector, 136.7 in the animal husbandry sector, 149.1 in the textile-clothing industry, 182.5 in the iron and steel industry, 1650.9 in the transportation sector, and 1985.9 in the energy sector (TÜROFED, 2010:7).

Tourism also facilitates investments in infrastructure, improves the quality of services, improves the quality of life for residents in tourist destinations by improving the components of the local infrastructure, including water and sewage systems, roads, electricity, telephone, and public transportation networks (Akal, 2010:135). Moreover, owing to tourism, the income and employment levels of the underdeveloped regions in a country are increased considerably, thus helping reduce the differences in regional development (Kantarcı and Karakaya, 2016:4). The regional distribution of the world's total tourism revenue in 2018 is presented in Figure 2.1.

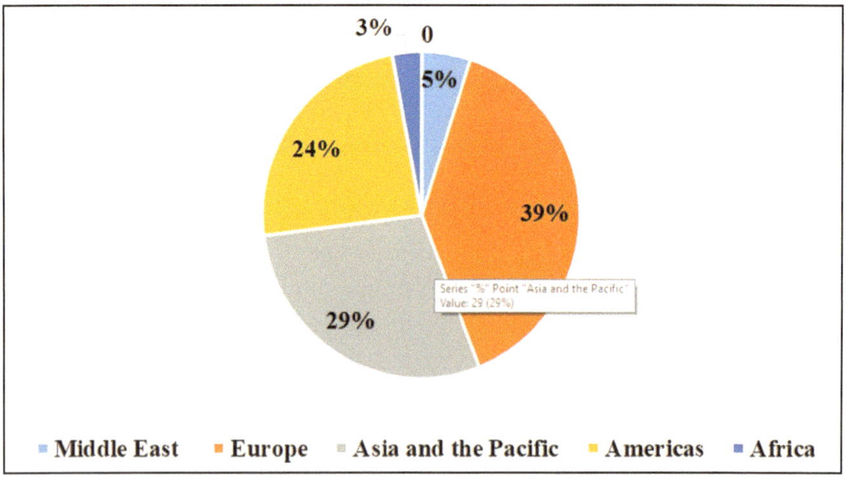

FIGURE 2.1 Continental distribution of world tourism revenue in 2017.
Source: UNWTO Tourism Highlights (2018).

It is seen that European countries took the largest share with 39% ($519 billion) of the world's tourism market, which constitutes around $1 trillion 340 billion. Asia-Pacific countries were in second place with 29% ($390 billion) of market share, and the Americas were in third place with a 24% share. Besides, the shares of African and Middle Eastern countries were 3% ($37 billion) and 5% ($68 billion), respectively.

Considering Figure 2.2, as of 2017, about 1 billion 326 million people as tourists seemed to be hosted around the world. It is observed that nearly 51% (672 million) of them chose to visit Europe, 24% (323 million) of them visited Asia-Pacific countries, 16% (211 million) of them visited the Americas, 5% (63 million) of them visited Africa, and 4% (58 million) of them visited Middle Eastern countries.

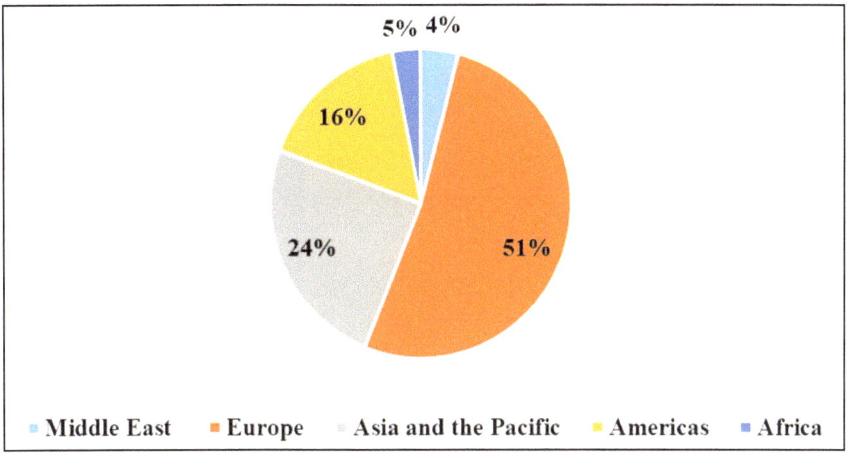

FIGURE 2.2 Continental distribution of tourist numbers around the world in 2017.
Source: UNWTO Tourism Highlights (2018).

2.3 THE PLACE OF TOURISM IN TURKISH ECONOMY

Tourism in Turkey started to show a significant improvement in the 1980s, and it has become one of the most important sources of income for the national economy. Until the 1980s, Turkey gave priority to import substitution policies, giving the utmost importance to industrialization in the manufacturing sector via state planning. Even though the manufacturing sector was heavily supported before the 1980s, the tourism industry was still expected to contribute to Turkish economy. During this period, tourism was regarded as a tool which created new employment opportunities, contributed to the balance of payments, and eased foreign exchange constraints (Gokovalı, 2010:142). In the period between 1984 and 1990, the growth of tourism reached around 12.1% in Turkey (Turizm Bakanlığı, 1998:57).

While the number of tourists visiting Turkey in 1980 was 1,288,060, this figure reached 46,112,592 in 2018, an approximately 30-fold increase. In terms of tourism income, the situation has not been different. Over the years, considering the data of UNWTO as well as those of the Culture and Tourism Ministry, Turkey has been one of the countries that have achieved significant growth in terms of the number of international visitors and tourism revenues. Table 2.1 shows the number of international visitors, tourism revenue, and average spending of visitors to Turkey. In 2018, visitors spent an average of $647 per day. Despite the increase in the number of tourists and tourism revenues, the average amount spent by tourists remains low.

TABLE 2.1 Turkey's Figures of Tourism Revenues, Number of Tourists, and Average Expenditures.

Year	Tourism income ($1000)	The number of tourists	The average spending by tourists ($) per day
1980	326,654	1,288,060	253.6
1990	3,225,000	5,389,308	621.3
2000	7,636,000	10,428,153	764.3
2010	24,930,996	33,027,943	755
2011	28,115,694	36,151,328	778
2012	29,007,003	36,463,921	795
2013	32,308,991	39,226,226	824
2014	34,305,904	41,415,070	828
2015	31,464,777	41,617,530	756
2016	22,107,440	31,365,330	705
2017	26,283,656	38,620,346	681
2018	29,512,926	46,112,592	647

Source: Ministry of Culture and Tourism (2019), http://yigm.kulturturizm.gov.tr/TR,72942/turizm-gelir-gider-ve-ortalama-harcama.html.

The direct contribution of tourism to the economy has been generally evaluated in terms of the percentage of the tourism revenue in the GDP and the percentage of employment provided by the tourism sector (Turguttopbas, 2019:3). Travel and Tourism directly supported 1.8% (495,000 jobs) of total employment in 2016. The total contribution of travel and tourism to employment, including jobs indirectly supported by the industry, was 8.1% of the total employment (2,197,500 jobs). Tourism continues to be an important economic sector capable of creating employment opportunities in Turkey. The share of tourism in Turkey was about 64% of the service revenues in the year 2017. The direct share of tourism revenues in the GDP was 4.1% in 2016, and its total share was 12.5% (http://www.kalkinma.gov.tr and https://www.wttc.org). Despite the tourist potential of Turkey, its market share from the global perspective is low. Constituting one of the focal points of world tourism, the challenge facing Turkey is the development of this potential in a continually changing market.

Turkey's geographical locations constitute a natural bridge between the old world continents of Asia, Europe, and Africa. Thus, since ancient times, relations between the continents have taken place largely over the Anatolian lands. The Anatolian lands were constantly used for migrations, wars, invasions, trade routes, and communication, and they have become the cradle of many civilizations. The union of all these civilizations formed the basis of today's Anatolian culture (UNWTO, 2014). In this context, the aim of Turkish tourism, being one of the focal points of world tourism, should be to increase its share in the growing world tourism market.

2.4 CONCLUSION

The tourism sector is a very important sector for many countries in the world, especially for underdeveloped and developing countries. For underdeveloped and developing countries, the tourism sector is one of the most important sources of employment creation and foreign exchange income. World Trade Organization (WTO) emphasizes that tourism is one of the most important dynamic economic activities nowadays and draws attention to the strength of this sector and its importance in the struggle against poverty (Bahar, 2008).

In this study, based on secondary data, the tourism sector in Turkey has shown significant progress since the 1980s and had a significant share in eliminating foreign deficits, reducing unemployment, and improving the balance of payments by providing a large foreign exchange inflow during periods of economic difficulties. Turkey is the sixth most popular tourist destination in the world (2015), attracting more than 30 million tourists annually and continuing to show positive growth year-on-year.

The tourism sector has set a target of 50 million tourist arrivals and $50 billion of tourism revenues by 2023. 30.9 million tourists visited Turkey in 2016 and generated approximately $18.7 billion of input to the economy.

Turkey has a significant and diverse tourism potential which remains underdeveloped. Consequently, Turkey should support domestic and foreign investments in the tourism sector, because of its positive spillover effects and higher returns to the economy. Turkey should continue to match tourism's demand for goods and services domestically to keep tourism revenues in the country and local communities to enhance the multiplier effect or to increase domestically added value in the tourism industry. There are insufficiencies in tourism facilities that provide an alternative to sun-sea-coastal tourism, which must be met to realize the countrywide tourism potential of Turkey. To move toward a better-balanced development of tourism, and thus, to benefit more people through tourism-based economic development, the government could support alternative tourism investments in the interior parts of Anatolia as soon as possible (Akal, 2009:149). Currently, Germany, the United Kingdom, and Russia are among the countries that visited Turkey the most. These tourists mainly visit Turkey for "sea, sand, and sun" holidays. However, Turkey has a number of natural beauties to offer various alternative tourism activities.

Development of tourism policies that take into account diverse elements such as the emerging Asian markets, development of alternative tourism, and creation of behavior profiles of potential tourists should provide important advantages to Turkey's tourism economy. For example, when we look at China, this country's GDP has been on a steadily increasing trend over the years. By 2020, its GDP is predicted to be around $21 trillion. In parallel with this development, personal tourism spending will continue to show a year-on-year increase. In this context, Turkey should increase its focus on the growing Chinese tourism market (Ongan et al., 2010).

As a result, the current data showing the number of tourists coming to the country, income earned from tourists, amount of expenditure per tourist and the share of tourism in the GDP, investments, and exports indicate that tourism is one of the leading sectors in Turkish Economy. Turkey's tourism activities in the world, in Europe and particularly in the Mediterranean region, are increasing at a fast pace. Although in the past, Turkey was not capable of competing with other countries, currently Turkey is a zealous country that is competing for side by side with these countries within every field of the tourism sector.

KEYWORDS

- **tourism**
- **tourism economy**
- **tourism sectors**
- **Turkey**

REFERENCES

Akal, M. Economic Implications of International Tourism on Turkish Economy. *Tourismos* **2010,** *5* (1), 131–152.

Bahar, O.; Kozak, M. Advancing Destination Competitiveness Research: Comparison between Tourists and Service Providers. *J. Travel Tourism Market.* **2007,** *22* (2), 61–71.

Bahar, O.; Kozak, M. *Tourism Economics Concepts and Practices*; Nova Science Publishers, Inc.: New York, 2008.

Çımat, A.; Bahar, O. Turizm Sektörünün Türkiye Ekonomisi İçindeki Yeri ve Önemi Üzerine Bir Değerlendirme. *Akdeniz İ.İ.B.F. Dergisi* **2003,** *6*, 1–18.

Giles, A.R.; Perry, A.H. The Use of a Temporal Analogue to Investigate the Possible Impact of Projected Global Warming on the UK Tourist Industry. *Tour. Manage.* **1998,** *19* (1), 75–80.

Gökovalı, U. Contribution of Tourism to Economic Growth in Turkey. Anatolia. *Int. J. Tour. Hosp. Res.* **2010,** *21* (1), 39–153.

Kantarcı, B. K.; Karakaya, U. In *The Effects of Tourism Sector on Turkey's Economic Development*, The 2016 WEI International Academic Conference Proceedings, Prague, Czech. https://www.westeastinstitute.com/wp-content/uploads/2016/07/kantarci-_-Karakaya-kocaeli-uni.pdf.

Ministry of Culture and Tourism. *Türk Turizminin Ekonomi ve Sosyal Yaşam İçindeki Yeri.* 1. Turizm Şurası: Ankara, 1998.

Ministry of Culture and Tourism. *Tourism Strategy of Turkey 2023.* Ankara, 2007.

Ongan, S.; Meng, F.; Mavnacıoğlu, K.; Dilbaz, N.; Bolcan, A.; Kırık, B. *Gelişen Çin Turizm Pazarı ve Türkiye*; İstanbul Ticaret Odası Yayınları, 2010; p. 36.

Turkish Hotelier Federation-TUROFED. *Tourism Report* 2010, No.: 2. Retrieved: http://www.turofed. org.tr/galeri/ cop/turi2.pdf.

Travel and Tourism Economic Impact 2018 Turkey. Available at: https://www.wttc.org//media/files/reports/economic-impact-research/countries-2018/turkey2018.pdf (accessed July 30, 2018).

Turguttopbaş, N. The Funding Structure of Turkish Tourism Sector and a Model Proposal for Tourism Revenue. *Int. J. Health Manage. Tour.* **2019,** (4) *1*, 1–15. DOI: 10.31201/ijhmt.528324

UNWTO Tourism Highlights, 2018. Available at: www.unwto.org/doi/pdf/10.18111/9789284419029 (accessed July 15, 2019).
UNWTO Silk Road Action Plan, 2014/2015; Madrid.
Ünlüönen, K.; Tayfun, A.; Kılıçlar, A. *Turizm Ekonomisi*.Nobel Yayıncılık: Ankara, 2015.
WTTC(WorldTravelandTourismCouncil).*TravelTourismPowerPerformanceReport*,2018. https://www.wttc.org/-/media/files/reports/2018/power-and-performance-rankings-2018.pdf.
http://www.kalkinma.gov.tr.

CHAPTER 3

CRITICAL REVIEW OF THE TOURISM PLANNING HISTORY OF TURKEY

UĞUR ÇALIŞKAN

Faculty of Tourism, Mugla Sıtkı Koçman University, Turkey

E-mail: ugurcaliskan@mu.edu.tr

ABSTRACT

Since tourism is a very comprehensive sector and is heavily affected by the current economic and political fluctuations, it is very difficult to plan. In this chapter, the evolution of tourism in the state structure and history of tourism planning, in Turkey, was discussed. Turkey hadn't had "particular" tourism policy document until 2007 and tourism planning was considered as a part of "Development Plans" which prioritized the economic benefits. In the first years, investments were envisaged by the public sector, and after 1980s, private sector investments were brought to the fore. In the period after year 2000, emphasis on diversification and quality in tourism has become increasingly apparent. On the other hand, although, in the first years, all the decisions were taken from the center and there was no clue how to implement, today regional priorities have come to the fore with the participative planning approach. Therefore, participatory planning approach is still "new" in Turkey, though, it can be stated that, in the future, tourism plans would be more participatory and sustainable.

3.1 INTRODUCTION

Planning, in its simplest definition, is the process of making the most appropriate choice among the alternatives that lead to the achievement of a desired goal (and therefore conducting the research and analysis necessary to determine these alternatives) and taking measures to mitigate risks and to identify the means of achieving the desired objectives (Murphy, 1983; Koçel, 2003).

The complex structure of the tourism sector, which includes not only many interdependent subsectors (accommodation, food and beverage, travel, souvenirs, etc.) at a national level but also international and external factors makes it difficult to plan. As in any business, tourism activities are subject to unexpected fluctuations. Therefore, the impact of cyclical movements should always be taken into consideration in tourism planning activities. The importance of tourism planning was revealed when the negative environmental, social, and cultural consequences of the rapid development of tourism with the emergence of mass tourism became apparent (Burns, 2004). However, tourism planning is not an old concept. Tourism researchers and decision-makers became interested in tourism planning in the 1970s (Alipour, 1996).

In many countries or regions, even though the concept of tourism planning is emerging as a new concept, it is widely accepted that tourism development should be planned (UNWTO, 1994). Such a perception does not mean that tourism planning is always done correctly and that it results in positive outcomes. Tourism planning was initially focused on achieving economic benefits, and environmental and social issues were considered secondary and even ignored (Mathieson and Wall, 1982; Getz, 1986; Archer and Cooper, 1998).

Beyond economic benefits, in order to increase the contribution of tourism to the development of a region and the country, it is also important to protect local cultural values and to reduce negative environmental effects. In addition, even if the plans considering only economic concerns have positive results in the short term, socioeconomic and environmental negative effects have emerged in the long term, and these problems have caused even the economic benefits to be unsustainable (UNWTO, 1994). In addition, within a purely economic perspective, it is not possible to determine appropriate development limits and to decide the right and appropriate strategies (Çalışkan, 2003).

Issues such as the general positive developments in world politics, legal, political, and technological developments easing travel, changes to current tourism demand toward differentiated products and the continuous emergence of new destinations (Goymen, 2000) underline the necessity of tourism planning. Apart from the increased competition that emerged with these developments, planning is necessary to minimize the negative effects while achieving the desired positive effects.

The growth of the tourism sector in Turkey began in the 1980s and it is now one of the top ten destinations in the world tourism sector; however, tourism planning has evolved over time, and it has changed in the context of world tourism and planning paradigms. In this section, the changes in the development process and planning perspective in Turkish tourism will be discussed. I will first present a brief overview of both state and private administrative development of the sector, followed by an examination of the tourism perspective of national and regional plans from the 1960s onwards (the planned period) and finally, contemporary planning approaches will be introduced in a critical manner.

3.2 THE DEVELOPMENT AND PLANNING OF TOURISM IN TURKEY

The formal emergence of tourism, in 1863, occurs with the "Ottoman in General" exhibition (Sergi-i Umumi Osmani) which opened in that year. The exhibition aroused great interest in Europe and brought groups from various European cities to İstanbul. An interesting coincidence is that the first organized tour to Europe from the Ottomans was in the same year (Akçura, 2013). "Instructions on Hotels and Guest Houses in and around Istanbul" (Dersaadet ve Bilâd-ı Selâse'de Bulunan Otel ve Misafirhaneler Hakkında Talimat), published in 1895, was an important legal regulation to control and regulate accommodation activities (Evcin, 2016). Although in the last period of the Ottomans, there were some legal regulations; tourism started to be significant in socioeconomic life only after the establishment of the Republic of Turkey (Evcin, 2016).

3.2 1 HISTORY OF TOURISM IN PUBLIC ADMINISTRATION IN TURKEY

In 1923, the Touring and Automobile Association of Turkey was founded and it provided services such as publishing tourism materials, training Turkish tourist guides, and promoting Turkey in the absence of a state tourism organization (Akçura, 2013). Following the establishment of the Republic, special importance was given to promoting the country. In addition to hosting international tourism meetings in 1930–1931, public institutions tried to promote the country by participating in fairs in France, Germany, Italy, Greece, and Serbia between 1932 and 1937 (Evcin, 2016). Legislation setting standards for tourist guides have also been constituted since the early years of the Republic (Evcin, 2016).

The General Directorate of Highways and the Republic of Turkey State Railways played significant roles in accelerating travel and tourism activities in the 1920s as highways and railways had become the main means of transportation (Evcin, 2016). In addition, through French, Italian, German, and American airlines, both domestic and international air transport began (As, 2013). Connecting flights were important to connect Turkey to the world (Evcin, 2016). Turkey's first major travel agency was National Turkish Tourist Agency (Natta) founded in 1925. Natta was also the Turkey representative for many international organizations. There were also travel agencies such as Pasrapid, Turkish Traveling & Tourist Agency (Tutta), Le Globe, Buleks (Istanbul Ekspres), Istanbul Travel Agency (ITA), which were established in the same period as Natta and generally dealt only with ticket sales (Akçura, 2013).

In Turkey's public administration, tourism started its journey in 1934 as "The Tourism Bureau" within the Ministry of Economy and continued as "The Tourism Branch" in 1939, within the Ministry of Commerce (Ayözcan and Önen, 1995). Considering the duties and fields of the ministries they were affiliated with, "The Bureau" and "The Branch" offices carried out propaganda for the regime rather than tourism (Andaç, 2004). In 1939, as a positive development, an Interministerial Tourism Commission was established to ensure coordination of the work of the institutions (TÜRSAB, 2018). Tourism-related activities were included in the responsibilities of the General Directorate of Press and Publications in 1940, and the Tourism Branch was renamed the "Department of Tourism" in 1943 (Andaç, 2004; İnan, 2009). The "Tourism Advisory Board" was established within the General Directorate of Press and Tourism established in 1949 by Law No. 5392 and the first "Tourism Advisory Meeting" was held. The meeting led to the preparation of a "Tourism Master Program" which formed the basis of tourism policies (Usta, 2001) and policies to support the private tourism sector (TÜRSAB, 2018). The "Tourism Establishments Incentives Act" (law number 5647), "Tourism Industry Incentives Law" (law number 6086) which repealed the previous one and the "Law on Foreigners' travels and Residence in Turkey" (law number 5683) can be noted as the first comprehensive legal regulations of tourism. The main purpose of these regulations were to increase the economic benefit of tourism by facilitating the travel of tourists within the country and establishing a financial aid system to reduce the start-up costs of tourism enterprises (Ayözcan and Önen, 1995). In the period up to 1950, due to the conjuncture caused by World War II, the organizations listed above could not deal with tourism sufficiently. Only the works of the Tourism Advisory Board were efficient. The Board pioneered a better understanding of the importance of tourism and the country's tourism potential (Ayözcan and Önen, 1995).

The structure established in 1950 as the "Tourism Credit Fund" was converted to the "Tourism Bank of Turkey" to achieve the objectives of Tourism Industry Incentives Law. Until it was incorporated with the Development Bank of Turkey in 1988, the Tourism Bank made huge contributions to the development of tourism in Turkey by building accommodation facilities in addition to providing loans with favorable terms and maturity to the private sector (Ayözcan and Önen, 1995; Andaç, 2004). It provided loans totaling USD 450 million (calculated at annual official exchange rates) for a total capacity of 146,700 beds. The Tourism Bank also built and operated accommodation facilities to act as "models" for the private sector (Goymen, 2000). The accommodation facilities operated under the TURBAN brand were privatized gradually in the 1980s and 1990s.

In 1957, the Ministry of Press and Tourism was established, and in 1963, the Directorate General of Press and Publications which was responsible for the press-related works of the ministry was transferred to the office of the Prime Minister and the name of the ministry became Ministry of Tourism and Promotion (Andaç, 2004). So, the journey of tourism in the public administration of Turkey which started in 1934 as a bureau finalized as a Ministry in 1963. Having merged with the Ministry of Culture in 1983, the name of the ministry became the Ministry of Culture and Tourism. Even though the ministry was divided into "Ministry of Tourism" and "Ministry of Culture" in 1989, the two ministries were reunited again with Law No. 4848 on the "Organization and Duties of the Ministry of Culture and Tourism", which was enacted in 2004, and has operated as the Ministry of Culture and Tourism since then.

Beyond these public bodies, with the growth of the tourism sector, nongovernmental organizations consisting of actors in the sector have emerged. In 1972, the Association of Turkish Travel Agencies (TÜRSAB) was established to promote and maintain professional ethics, to protect tourists, and to cooperate with the Ministry of Tourism in marketing activities (Goymen, 2000). The Association of Tourist Guides (TUREB) was established in 2012 with the goals of tourist guides to play a more effective role in the country's tourism, to direct them to operate in line with the country's culture and tourism policies, and to prevent illegal tourist guidance activities (TUREB, 2013). Both of these two organizations serve as public institutions.

The Tourism Development and Education Foundation (TUGEV) was established in 1985 to conduct research on tourism, to contribute to the development of training programs and to improve existing ones, and to cooperate with national and international organizations (Goymen, 2000); the Turkish Hoteliers Association (TUROB) was established in 1995 to regulate professional standards and ethics, to facilitate price stability, to provide general and on-the-job training, to promote and market destinations, and to cooperate with national and international tourism organizations (TUROB, 1997). The Turkish Tourism Investors Association (TTYD) was founded in 1988 to bring together investors, to share information and coordinate their activities, to provide technical assistance, and to act as a lobby group (Goymen, 2000).

3.2.2 TOURISM PLANNING IN TURKEY

The year 1960, when the State Planning Organization (SPO) was founded and 5-year development plans were considered to be the main instrument of state programs, is accepted as the beginning of the "Planned Period of Turkey". The tourism sector has been developed under expectations of economic benefits within the framework of the 5-year development plans and, even if small, tourism has found a place in development plans and some basic national decisions have been taken. Development plans are mandatory for the public sector and a lodestar for the private sector (Uca, 2018), and have been prepared with a "top-down" perspective with rational planning logic and without focusing on actions and only considering the financial element. It would be appropriate to state that the development plans are designed with the view that tourism can be improved with a series of economic-based public incentives such as the allocation of state-owned resources or the provision of loans to enterprises (Tosun and Jenkins, 1996).

In terms of tourism, the planned period can be analyzed in three subperiods; the 1963–1982 period when state-led policies and public bodies were dominant; the 1982–2003 period which focused on private investments; and the period of tourism diversification after 2004 (Table 3.1).

TABLE 3.1 Periods 5-Year Development Plans.

	Period
First 5-Year Development Plan	1963–1967
Second 5-Year Development Plan	1968–1972
Third 5-Year Development Plan	1973–1977
Fourth 5-Year Development Plan	1979–1983
Fifth 5-Year Development Plan	1985–1989
Sixth 5-Year Development Plan	1990–1994
Seventh 5-Year Development Plan	1996–2000
Eighth 5-Year Development Plan	2001–2006
Ninth Development Plan	2007–2013
Tenth Development Plan	2014–2018
Eleventh Development Plan	2019–2023

The first period (1963–1982) is characterized by pioneering infrastructure projects for tourism development (South Antalya tourism development project, Side project, etc.), and the standardization and certification of tourism accommodation and daily catering facilities.

The First, Second, Third, and Fourth Development Plans which cover the years between 1963 and 1982 adopted the principle of the intensification of investments in priority regions, arguing that if tourism activities were developed throughout the whole of Turkey, the infrastructure and superstructure would be inefficient and the investments would be unproductive. In this period, when the public sector pioneered and tried to prepare the basic conditions for development, tourism policies emphasized focusing on mass tourism, the long-term allocation of forest land for accommodation facilities (SPO, 1972), and tourism investments were steered to Aegean, Antalya, and Marmara regions. The infrastructure of these regions, such as airports, highways, water, electricity, sewerage facilities, and marinas was built by the state (SPO, 1972). Consequently, while tourism developed in the coastal regions, the inner regions were ignored and a spatial segregation emerged (Goymen, 2000). The facilities (including accommodation ones) were built by the public, financial, and legal measures were imposed and spatial plans were approved to support private sector investments.

In all plans, not only the capacity of accommodation facilities, but also the need to eliminate infrastructure deficiencies in tourism zones were emphasized. Shortages of qualified personnel were also mentioned frequently and the development plans emphasized the need for the training of personnel. In the Second Development Plan, it was decided to establish the "National Center for Tourism Education and Training" to carry out tourism education and training activities in a holistic way. In the Fourth Development Plan, the Ministry of Tourism and Promotion was assigned the responsibility for coordinating the training and employment of tourism personnel. Although it is not mentioned in the following plans, the first plan proposed to establish a Tourism Development Board with the participation of relevant institutions and organizations at town and city level. However, these boards were never established.

The SPO and the Ministry of Tourism, in the 1960s, prepared nation-wide master tourism plans and in 1969, the coastal zone between the southern boundary of Çanakkale (Dardanelles) and the western boundary of İçel (Mersin) were identified as a priority area for tourism development and a Tourism Etude for Western Turkey was undertaken (Ayözcan and Önen, 1995). In the 1970s, a Planning Department was established in the Ministry of Tourism and Promotion which aimed to expand the etude to the whole country. The department completed and approved the 1/25.000 scale spatial plans in the Southwest Anatolia and Cappadocia Regions in 1971 (Ministry of Tourism and Culture, n.d). Nevertheless, it is not possible to say that the 70s were productive in terms of tourism due to political instability and economic crises (Akçura, 2013).

The year 1982 when the Tourism Incentive Law, which gives more authority to the Ministry of Tourism came into force is considered the beginning of the second period in tourism development. The Tourism Incentive Law sought to overcome obstacles such as the "lack of coordination among various tourism-related organizations, the difficulty of obtaining land suitable for investments, and discouragingly long formalities" (Göymen, 2000, p 1033). The law had provisions for the long-term allocation of publicly owned land to investors, the provision of the main infrastructure by the government; long, medium, and short-term loan support for construction, furniture, and operational costs and additionally, in priority areas, provided further incentives for tourism enterprises via subsidized utility prices in electricity, water, and gas consumption (Alipour, 1996; Uca, 2018). Under the new law, the formalities and conditions for the allocation of state-owned land were simplified along with a reduction in the number of state institutions involved and tourism investments were directed to priority regions as indicated in the Fourth Development Plan (Goymen, 2000). After 1983, state investment in pioneering accommodation facilities ceased and preparations for the privatization of existing institutions began. The fixing of tourism facilities' prices by the government ceased, thus enabling market mechanisms to operate. A detailed system of incentives using the Ministry of Tourism and the Tourism Bank as the main delivery channels efficiently supported the tourism sector at least until the end of the 80s (Goymen, 2000).

While the emphasis on mass tourism continued in the plans of this period, the development of other tourism types (winter, hunting, thermal, festival, health, youth, entertainment, eco, plateau, golf, caravan, congress, cruise, 3rd age tourism, etc.) were mentioned in the fifth plan and afterwards. In line with the new trends in tourism demand, the development of small-scale enterprises and certification system were prioritized to improve service quality (SPO, 1995). However, the planning approach and practice did not change and the tourism sector was dealt with only from an economic perspective. There were not any fundamental changes in the determination of tourism regions either and the main criterion continued to be to maximize the number of tourists and foreign exchange revenue. Furthermore, the identification of priority regions reflected political concerns in some respects. The central government wanted to control tourism development throughout the country. Thus, such regionalization contributed to further centralization and threatened the sustainable development of tourism (Tosun and Jenkins, 1996). On the other hand, factors such as the Chernobyl nuclear accident in the Soviet Union in 1986, the 1991 Gulf War, and the 1999 Marmara Earthquake led to a decline in tourism demand for Turkey (Demir, 2004). Such incidents have led to changes in Turkey's tourism and "all-inclusive system" and "low-cost air transport" was prevalent in those years (Uca, 2018).

The 3rd period which can be identified as the tourism diversification period began in 2004. With the amendment made to the Tourism Incentive Law in 2003, the Ministry of Culture and Tourism was granted the authority to approve plans at all scales for cultural and tourism protection and development zones and tourism centers determined by the decision of the Council of Ministers. The policies of the Ministry of Culture and Tourism have also evolved leading to an increase in the diversity of tourism in the country and to spreading tourism beyond the coastal areas. In this period, besides the number of tourists, emphasis on quality and branding are brought to the forefront along with a goal of developing new types of tourism rather than solely focusing on coastal tourism. The priorities in tourism incentives was given to marketing, air transport, improving the quality of labor and service (SPO, 2000), attracting higher income groups (Ministry of Development, 2013b), the development of health tourism, and protecting the natural, historical, and social environments (SPO, 2006).

Sustainability is firstly mentioned in the Tenth Development Plan and the participation of local governments and stakeholders in tourism planning is mentioned in the Ninth Development Plan. However, this participation is limited to "the tourism-intensive regions" and to finance infrastructure investments (Ministry of Development, 2013a).

After the 2000s, the policies of the Ministry of Culture and Tourism have also evolved to increase the diversity of tourism in the country and to spread tourism beyond the coastal regions. With the amendment made to the Tourism Incentive Law in 2003, the authority to approve plans at all scales allowed the Ministry of Culture and Tourism to make holistic plans for all tourism regions. However, there has been no fundamental change in the planning paradigm. Local participation is at the level of "pseudo participation" and is usually limited to meetings where information is given to local people about planning activities. Local people rarely have the chance to convey their opinions and suggestions, and even in such cases, decisions are taken by the public authorities and planners. In addition, the governments have continued to act as reactive (trying to solve problems) rather than being proactive (taking preventive measures) (Goymen, 2000).

The preparation and approval of the Tourism Strategy of Turkey (TST) in 2007 was one the most important developments of this period. With the document in question, a national policy for the tourism sector, which by then had reached a significant position in the Turkish economy, was developed. The TST was prepared in line with the governance principle and to ensure the cooperation between public and private sectors in the administration and implementation of strategic planning activities (Ministry of Tourism and Culture, n.d). The TST claims to promote tourism "not on an individual land plot but on a regional basis; with a guiding and dynamic rather than imperative and static" framework (Ministry of Culture and Tourism, 2007, p 1). It is the first guiding document beyond the development plans and the national tourism vision and general objectives have been determined. The main targets of the TST are to reach 63 million tourists, $ 86 billion in foreign tourism revenue and $ 1350 expenditure per tourist in 2023. The vision of TST is that "with the adoption of sustainable tourism approach, tourism and travel industry will be brought to a leading position for leveraging rates of employment and regional development and it will be ensured that Turkey becomes a world brand in tourism and a major destination in the list of the top five countries receiving the highest number of tourist and highest tourism revenues by 2023" (Ministry of Culture and Tourism, 2007, p 3).

The aim of the TST is to plan the areas with high potential as corridors and to develop a "land development model" that encourages private sector investments. In this way, it is aimed to utilize the idle thermal, winter, and nature, etc. potential of these areas (Ministry of Tourism and Culture, n.d) (Figure 3.1).

One of the most significant changes in terms of planning practices in Turkey has been realized during the integration process to the European Union. Following the adoption of Nomenclature of Territorial Units for Statistics (NUTS), Turkey is divided into 26 Level II units and in accordance with the Law No. 5449, regional development agencies were established in each region (Table 3.2). Regional plans, which aim to integrate national-scale plans and programs with regional priorities and actions and to increase local participation and responsibility, have been prepared by development agencies which are also local institutions (General Directorate of Legislation Development and Publication, 2018).

The preparation process of regional plans by development agencies was a new planning experience in Turkey. Though efficiency and equity of participation varied from region to region, the plans have been prepared with the participation and guidance of actors from the public sector, civil society, the private sector, university, local people, and national stakeholders. While this process was described as participative, it is better understood as consultative, that is, an opportunity was given to non-state actors to express their views and opinions. However, there was no requirement that these views be taken into account when the plans were produced

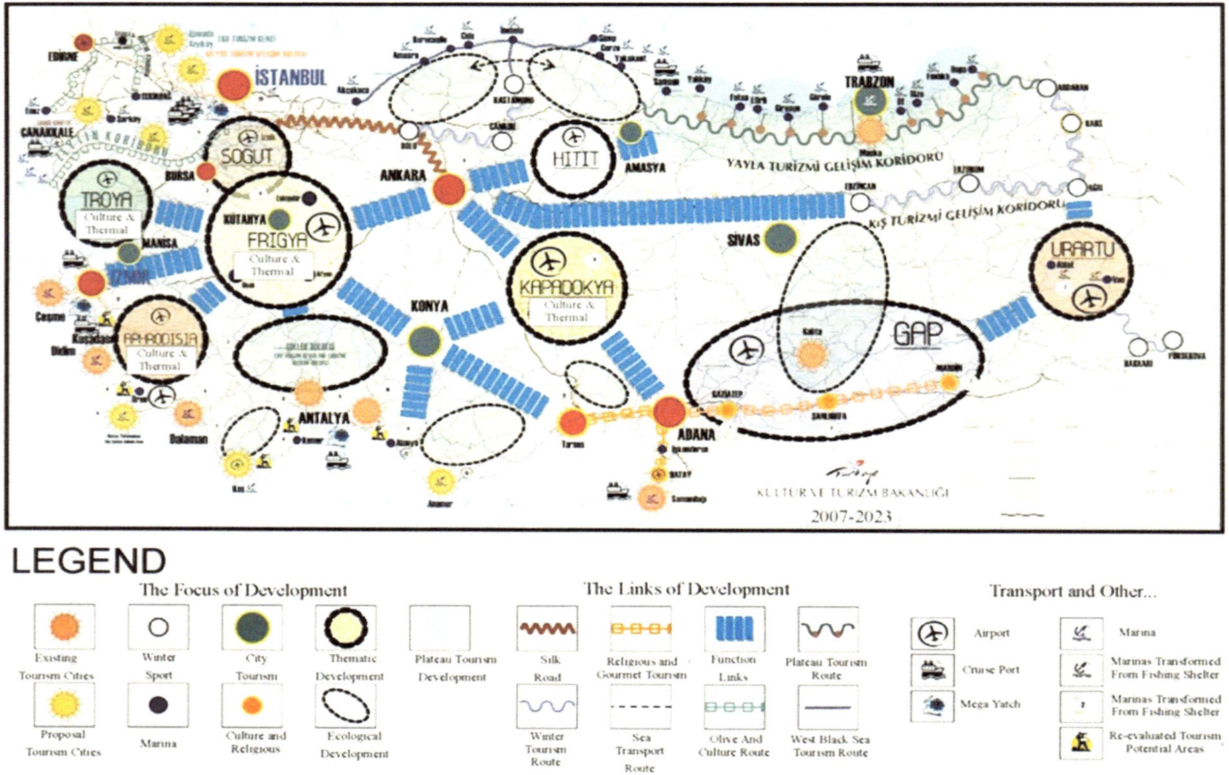

FIGURE 3.1 Tourism strategy of Turkey conceptual plan.
Source: Ministry of Culture and Tourism, 2007.

by the regional development agencies. A truly participative process would have involved all the actors in all stages of the planning exercise.

TABLE 3.2 Regional Development Agencies in Turkey and Provinces.

Development Agency	Provinces[a]
İstanbul Development Agency	İstanbul
Trakya Development Agency	Tekirdağ, Edirne, Kırklareli
South Marmara Development Agency	Balıkesir, Çanakkale
İzmir Development Agency	İzmir
Southern Aegean Development Agency	Denizli, Aydın, Muğla
Zafer Development Agency	Kütahya, Afyonkarahisar, Manisa, Uşak
Bursa, Eskişehir, Bilecik Development Agency	Bursa, Bilecik, Eskişehir
East Marmara Development Agency	Kocaeli, Bolu, Düzce, Sakarya, Yalova
Ankara Development Agency	Ankara
Mevlana Development Agency	Konya, Karaman
West Mediterranean Development Agency	Isparta, Antalya, Burdur
Çukurova Development Agency	Adana, Mersin
Eastern Mediterranean Development Agency	Hatay, Kahramanmaraş, Osmaniye
Ahiler Development Agency	Nevşehir, Aksaray, Kırıkkale, Kırşehir, Niğde
Central Anatolia Development Agency	Kayseri, Sivas, Yozgat
Western Black Sea Development Agency	Zonguldak, Bartın, Karabük
North Anatolian Development Agency	Kastamonu, Çankırı, Sinop
Central Black Sea Development Agency	Samsun, Amasya, Çorum, Tokat
Eastern Black Sea Development Agency	Trabzon, Artvin, Giresun, Gümüşhane, Ordu, Rize
Northeast Anatolia Development Agency	Erzurum, Erzincan, Bayburt
Serhat Development Agency	Kars, Ağrı, Ardahan, Iğdır
Fırat Development Agency	Malatya, Bingöl, Elazığ, Tunceli
Eastern Anatolia Development Agency	Van, Hakkâri, Bitlis, Muş
Silk Road Development Agency	Gaziantep, Adıyaman, Kilis
Karacadağ Development Agency	Şanlıurfa, Diyarbakır
Tigris Development Agency	Mardin, Batman, Şırnak, Siirt

Source: Ministry of Foreign Affairs, Directorate of European Union (2014).

[a]The first province indicates where headquarter of the development agency is located. Other provinces are arranged in alphabetical order.

Although there are minor variations between regions, tourism is generally stated as an important sector in all the regional plans, and tourism is especially emphasized in the Eastern Black Sea, North Anatolia, and South East plans except for the regions where tourism has already developed (Çalışkan and Toy, 2016). The Level II Regional Plans mainly focused on:

- Infrastructure works,
- Increasing human capital,
- Improving environmental infrastructure,
- Improving the quality of life of both tourists and residents in urban and rural areas,
- Increasing the capacities and service quality of the tourism enterprises,
- Eliminating the shortage of qualified personnel,
- Preservation and development of natural, historical, and sociocultural values (to be transformed into tourism products),
- To increase tourism and product diversity,
- Raising the local awareness, and
- Promotion activities

These measures show that in terms of overlapping of the local response to the national targets and problems, the national and local policies are in harmony.

Currently, the Eleventh Plan is in force in Turkey as a national scale plan. In the plan approved by the Grand National Assembly in July 2019, tourism along with agriculture and defence industry, were given "priority development sector" status. At the end of the plan period, the number of visitors is expected to reach 75 million, 67.7 million of which are international, and tourism revenue is expected to reach 65 billion USD. The figures indicate that the objectives of the TST have been revised and a more realistic approach is observed in tourism revenue while an increase in the number of tourists is expected. In order to achieve these goals, tourism needs to be diversified to include gastronomy, golf, cruise, weddings (especially for the Indian market), faith, congress and shopping, and especially health and sports tourism. Other goals are the improvement of service quality and the transformation of the sector. The social benefits of tourism are also underlined (Presidency of Turkey, Presidency of Strategy and Budget (SBB), 2019).

In the period of 2019–2023, it is stated that a "Tourism Law" would be put forth as a base legislation on all tourism-related issues and a "Tourism

Promotion and Development Agency" would be established to direct tourism promotion and investment in Turkey. Furthermore, Management Plans have to be developed to improve the perception and image of Turkey both in existing and new target markets, as well as Sport Tourism Strategy of Turkey to ensure the long-term development of sport tourism (SBB, 2019).

The holistic planning of tourism destinations mentioned in the TST but not implemented is also amongst the Eleventh Development Plan objectives. However, no change in the planning approach was emphasized and it was only stated that "sustainable tourism practices would be developed with an environmentally sensitive and responsible approach" (SBB, 2019, p 100).

Regarding transportation, an increase in the number and volume of direct flights from international markets to tourism destinations and, for the first time in the development plans, the tourism use of train voyages and routes was emphasized. It is also stated in the plan that R&D activities and technology-based implementations in tourism will be supported to benefit more from internet technology and from the sharing economy.

In this plan, although basically economic factors are taken into consideration, it is also to increase the benefits of tourism to rural development and society. Within this framework, there is a glimmer of hope that the first steps in a much needed transformation in the tourism planning paradigm and practice in Turkey have been made.

3.3 CONCLUSIONS

Planning is a future-focused effort. In this context, decisions need to be taken about many issues. However, old planning paradigms and approaches cannot respond to rapidly changing conditions of the present-day and new approaches should be developed. As Ersoy (2012) points out, the plans in the past had only targets, and no methods and/or paths were specified for how to achieve the targets. One of the main problems of tourism development in Turkey is a central planning approach which leads to decisions that do not take into account regional and local conditions (Tosun and Jenkins, 1996) and often results in the alienation of local populations in tourism development. The Ministry of Culture and Tourism and the institutions responsible for authorizing tourism investments and incentives are accessible to the wealthy and educated elite, but not to the majority of indigenous people living in tourist areas. In this sense, there has been a large communication gap between local communities and decision-makers (Tosun and Jenkins, 1996). Therefore, tourism development did not help to reduce the disparity between regions, but expanded it further. For example, 85% of tourism activities and 86% of tourism investments in are in Marmara, Aegean, and Mediterranean, the most developed regions of Turkey. These rates are 2.8% and 5.5%, in Eastern and Southeastern Anatolia, respectively, which are the least developed regions in Turkey (Tosun and Jenkins, 1996). Moreover, this central planning practice and the lack of a strategic plan has resulted in intense construction and pollution of resources and has endangered the future of tourism development in the Mediterranean and Aegean coasts due to incomplete and fragmentary land use policies (Alipour, 1996).

Another fundamental problem of the sustainable tourism development in Turkey is political instability. Unstable and incoherent policies have led to uncertainty and almost a laissez-faire approach to tourism development (Tosun and Jenkins, 1996). On the one hand, the passive role of the government in tourism and on the other hand, the lack of a proactive planning process led to the development of the industry by external actors in an almost uncontrolled way (Alipour, 1996).

The new paradigm, appropriate for the modern tourism development, should overcome the shortcomings of the traditional approach. With Butler's (1991) expression, this new synthesis depends on the coordination of policies, proactive planning, the adoption of restrictions on growth, training of all parties, the commitment to a long-term perspective, and the linking of tourism and sustainable development. Innovative planning should be action-oriented to ensure the optimal use of all resources through activating local government and provide creative solutions to corporate change and transformation in uncertain social situations and market changes (Alipour, 1996).

Regional Development Agencies have adopted a different practice compared to the centralized state structure and rational planning experiences in Turkey. In this framework, with increasing experience in and awareness of participatory planning processes in both public bureaucracy and local actors, it should be possible to produce plans better reflecting the needs and expectations of the populations. To achieve sustainable tourism development, a new paradigm is required. This new paradigm will raise the awareness and understanding of the contributions of tourism to society, the environment and the economy; it will ensure a wider distribution of the benefits of tourism development; emphasize the importance of the protection of the social and natural environments, and will focus on providing the highest level of customer satisfaction (Alipour, 1996).

KEYWORDS

- **tourism policy and planning**
- **rational planning**
- **planning period**
- **Turkey**

REFERENCES

Akçura, G. *Türkiye Turizminde 150 Yıl (150 Years of Tourism in Turkey)*; OTI Holding Yayınları: Antalya, 2013.

Alipour, H. Tourism Development Within Paradigms: The Case of Turkey. *Tour. Manag.* **1996,** *17*(5), 367–377.

Andaç, F. *Turizm Hukuku*; Detay Yayıncılık: Ankara, 2004.

Archer, B. H.; Cooper, C. The Positive and Negative Impacts of Tourism. In *Global Tourism: The Next Decade*, 2nd ed.; Theobald, W., Ed.; Butterworth-Heinemann: Oxford, 1998.

As, E. *Cumhuriyet Dönemi Ulaşım Politikaları (1923–1960).* AKDTYK Atatürk Araştırma Merkezi Yayınları: Ankara, 2013.

Ayözcan, A.; Önen, O. *Türkiye'de Turizm Sektöründeki Gelişmeler ve Türkiye Kalkınma Bankası'nın Rolü.* Kalkınma Bankası Yayınları: Ankara, 1995.

Burns, P. M. Tourism Planning, A Third Way. *Ann. Tour. Res.* **2004,** *31*(1), 24–43.

Butler, R. W. Tourism, Environment, and Sustainable Development. *Environ. Conserv.* **1991,** *18*(3), 201–209.

Çalışkan, U.; Toy, S. Kalkınma Ajanslarının Planlama ve Finansman Açısından Turizm Sektörüne Sağladığı Katkılar. *İşletme Fakültesi Dergisi* **2016,** *17*(2), 141–160.

Çalışkan, U. Turizmin Yerel Halk Üzerindeki Etkileri ve Kuşadası İlçesi Uygulaması, Master Thesis. Dokuz Eylül Üniversitesi, Sosyal Bilimler Enstitüsü, İzmir, 2003.

Demir, C. A Profile of Turkish Tourism. *Int. J. Contemp. Hosp. Manag.* **2004,** *16*(5), 325–328.

Ersoy, M. *Kentsel Planlama Kuramları;* İmge Kitabevi: Ankara, 2012.

Evcin, E. Türkiye Turizm Tarihinde Öncü Bir Eser: Turizm Kılavuzu 1934–1935. *Anatolia: Turizm Araştırmaları Dergisi* **2016,** *27*(1), 7–24.

General Directorate of Legislation Development and Publication. İmar Kanunu www.mevzuat.gov.tr/MevzuatMetin/1.5.3194.pdf (accessed March 22, 2019).

Göymen, K. Tourism and Governance in Turkey. *Ann. Tour. Res.* **2000,** *27*(4), 1025–1048.
İnan, B. Cumhuriyet Dönemi Türkiye Turizmi ve Bir Turizm Modeli Önerisi, Master Thesis, Dokuz Eylül Üniversitesi, Sosyal Bilimler Enstitüsü, İzmir, 2009.
Koçel, T. *İşletme Yöneticiliği;* Beta Basım Yayın: İstanbul, 2003.
Mathieson, G.; Wall, G. *Tourism: Economic, Physical and Social Impacts*; Longman: London, 1982.
Ministry of Culture and Tourism. Tourism Statistics. http://yigm.kulturturizm.gov.tr/TR-9851/turizm-istatistikleri.html (accessed July 13, 2019).
Ministry of Culture and Tourism. Tourism Strategy of Turkey 2023; Ankara, 2007. https://www.ktb.gov.tr/EN-138780/tourism-strategy-of-turkey.html (accessed June 13, 2019).
Ministry of Development. *9. Development Plan* (in Turkish); Ankara, 2013a.
Ministry of Development. *10. Development Plan* (in Turkish); Ankara, 2013b.
Ministry of Foreign Affairs, Directorate of European Union. Regional Development Agencies (in Turkish). http://www.yereldeab.org.tr/TabId/100/ArtMID/3294/ArticleID/1951/KalkinmaAjanslari.aspx (accessed May 24, 2019), 2014 Ministry of Tourism and Culture. *Tourism Planning Work Notes* n.d.
Murphy, P. Tourism as a Community Industry, an Ecological Model of Tourism Development. *Tour. Manag.* **1983,** *4*(3), 180–193.
Presidency of Turkey, Presidency of Strategy and Budget (SBB). Eleventh Development Plan. *www.sbb.gov.*tr/wp-content/uploads/2019/07/OnbirinciKalkinma Plani.pdf. (accessed 19 July, 2019), 2019.
SPO. *Third Development Plan (1973–1977)* (in Turkish); Ankara, 1972.
SPO. *Seventh Development Plan (1996–2000)* (in Turkish); Ankara, 1995
SPO. *Eighth Development Plan (2001–2005)* (in Turkish); Ankara, 2000.
SPO. *Ninth Development Plan (2007–2013)* (in Turkish); Ankara, 2006.
Tosun, C.; Jenkins, C. L. Regional Planning Approaches to Tourism Development: The Case of Turkey. *Tour. Manag.* **1996,** *17*(7), 519–531.
TUREB. Hakkımızda. http://tureb.org.tr/tr/Page/Detail/50 (accessed July 12, 2019).
TUROB. *Annual Report* (in Turkish) Mimeographed, 1997.
TÜRSAB. Dünden Bugüne Seyahat Acentaları. https://www.tursab.org.tr/tr/seyahat-acentalari/dunden-bugune-seyahat-acentalari_501.html (accessed March 21, 2019).
Uca, S. Türkiye Turizmine Genel Bakış. In *Türkiye Turizm Coğrafyası*; Duran, E., Sertkaya Doğan, Ö., Eds.; Paradigma Kitapevi: Çanakkale, 2018; pp 199–214.
UNWTO. *National and Regional Tourism Planning*; UNWTO: Madrid, 1994.
Usta, Ö. *Genel Turizm*; Anadolu Matbaacılık: İzmir, 2001.

CHAPTER 4

ALTERNATIVE TYPES OF TOURISM THAT CAN BE APPLIED FOR SUSTAINABLE TOURISM IN TURKEY

L. KARADAĞ

Muğla Sıtkı Koçman University, Datça Kazım Yılmaz Vocational School, Datça, Muğla

E-mail: leventk@mu.edu.tr

ABSTRACT

Turkey has put emphasis on the development of tourism especially after 1980 and has carried out various studies in this direction, especially in the Tourism Incentive law. These studies have presented positive results and there has been an increase in the number of tourists coming to Turkey every year and the income from tourism. Although not large enough in the world tourism market, Turkey has significant tourism statistics compared to rival countries. Many destinations are visited by millions of tourists each year. The most important point here is the type of tourism that the tourists visiting Turkey prefer. Tourism products based on the understanding of mass and classical tourism (sea-sand-sun) are known to be consumed. Because of the implementation of mass tourism, the negative economic, social and environmental effects of tourism are observed. Bed capacities, whose numbers have been increasing year by year, remain dormant at times other than during the summer months, when tourists arrive intensively. In this context, productivity remains weak. It is inevitable to turn to alternative tourism products so that the capacity in question can be used regularly and efficiently. The most effective solution for combating these problems is to ensure the sustainability of tourism and to diversify tourist products. The preferred tourism products on the sea-sand-sun axis need to be diversified. This situation is inevitable in terms of sustainable tourism sector in Turkey. Tourist demand varies from individual to individual and cannot be met with a single standard product. On the other hand, these evolving and changing demands open up new opportunities to develop new products to satisfy these new demands and expectations. At this point, the differentiation of the most preferred types of tourism (sea-sand-sun) should be abstained from and different tourism products should be presented. From this point of view, alternative tourism is a necessity and raises as an issue that needs to be addressed at this point.

4.1 INTRODUCTION

For developing countries such as Turkey, the tourism sector is very important. Turkey has put emphasis on the development of tourism especially after 1980 and has carried out various studies in this direction, especially in the Tourism Incentive Law. These studies have presented positive results and there has been an increase in the number of tourists coming to Turkey every year and the income from tourism. The importance of tourism for Turkey can clearly be seen when the number of guests and income from tourism status are examined below.

When Table 4.1 is examined, it is seen that Turkey's tourist numbers have been steadily increasing in the last 15 years. Especially during the last 5 years, nearly 40 million tourists preferred Turkey for their holidays. According to the World Tourism Organization, the number of tourists travelling around the world is 1.4 billion in 2018. In such a large consumer market, 40 million tourists do not seem as a great number, but compared to the rival countries (Greece, Spain, Italy, etc.) it is a considerable amount of tourists.

The most important point here is the type of tourism that the tourists visiting Turkey prefer. The preferred tourism products on the sea-sand-sun axis need to be diversified. This situation is inevitable in terms of sustainable tourism sector in Turkey. Alternative tourism raises as an issue that needs to be addressed at this point.

4.2 LITERATURE REVIEW

Today, sustainability has become an important phenomenon for continuously developing, growing, and consumption-oriented societies. This subject is studied in the international literature in different science fields (Cottrell et al., 2013; You and Zhang, 2017). Although the topic of sustainability was initially discussed with a focus on development, in time, it has started to be studied in different fields such as environment, agriculture, tourism, and economy (Yalçın, 2016; Caira and Ferranti, 2016; Novas et al., 2017; Çeken et al., 2007).

TABLE 4.1 Number of Foreign Visitors to Turkey and Income from Tourism.

Years	Incoming visitor	Tourism revenue	Average spending
2003	16,463,623	13,854,866	850
2004	20,753,734	17,076,607	843
2005	25,045,142	20,322,111	842
2006	23,924,023	18,593,951	803
2007	27,239,630	20,942,500	770
2008	31,137,774	25,415,067	820
2009	31,759,816	25,064,482	783
2010	32,997,308	24,930,997	755
2011	36,769,039	28,115,692	778
2012	37,715,225	29,007,003	795
2013	39,860,771	32,308,991	824
2014	41,627,246	34,305,903	828
2015	41,114,069	31,464,777	756
2016	30,906,680	22,107,440	705
2017	37,969,824	26,283,656	681
2018	46,112,592	29,512,926	647

The rapid consumption of natural, cultural, and economic resources will mean the end of the products produced with these resources and this might cause the tourism sector go down a road where there is no point return. Tourist product diversification is among the main building blocks of mutual win-win strategies for sustainable tourism. Thanks to the orientation of the tourism sector to different and new alternative forms of tourism from classical mass tourism, the problem of constantly consuming the same resources is relatively eliminated. Considering the resources that Turkey has, it will be seen that it is possible to create a wide variety of alternative tourism products that are different in quality and different from each other (Karadağ, 2016).

When tourism activities in the world are examined, the East has become a trend destination in the last decades. In addition, it is noted that the service sector has gained a significant increase in the share of production in the tourism sector. As a result, while tourism activities continue to develop and grow, they vary according to demands. With the changing demands, Turkey is becoming a center of attraction with its rich historical and cultural treasures in addition to its natural beauties (Ulusan and Batman, 2010).

Especially in countries such as Turkey, where tourism practices are often stuck on the sea-sand-sun axis, it has been observed that the concept of alternative tourism is developed in order to direct consumption to different channels and different tourism products are diversified via offering the products to tourists (Hacıoğlu and Avcıkurt, 2008).

According to Baykan (2014), this is a new trend because different touristic activities are being offered to tourists by incorporating new destinations into the tourism sector, providing environment-local people-tourist interaction and promoting activities throughout the whole year.

Different definitions of "alternative tourism" can be found in the literature. What these ideas have in common is that alternative tourism is a tourism idea created in response to mass tourism. However, alternative tourism is also expressed as a new developing understanding which aims to protect the environment and local people from the sustainability perspective (Arslaner and Erol, 2017).

In the tourism sector, which is considered as one of the biggest service branches in Turkey and in the world, demand is constantly increasing. As a result, it starts to bring diversity in supply activities with it. Considering that all the resources of a region (especially sea, sand, and solar resources) are scarce and depletable, it is imperative to diversify tourist products by changing the goals or wishes that lead tourists to travel (Karadağ, 2016).

Tourist demand varies from individual to individual and cannot be met with a single standard product. On the other hand, these evolving and changing demands open up new opportunities to develop new products to satisfy these new demands and expectations. At this point, the differentiation of the most preferred types of tourism (sea-sand-sun) should be abstained from and different tourism products should be presented. From this point of view, alternative tourism is a necessity (Özdemir, 2003).

Alternative tourism is best understood as response to the development of mass coastal tourism and any tourist activity that is not based upon the classic sun, sea, and sand triangle can be regarded as alternative tourism. In the literature, many different categories of alternative tourism products are presented. These categories are not mutually exclusive; some of them are broad while others are quite narrow. It is possible to come across the types of tourism that are intertwined with each other. The most important reason for this is that the types of tourism have close interest in each other and cannot be separated at a certain point. Alternative tourism products to be offered in Turkey are listed as follows (Çelik, 2018; Arslaner and Erol, 2017; Öztürk and Yazıcıoğlu, 2002; Çetinel, 2001, p 398; Kozak et al., 2001):

- Rural Tourism
- Farm Tourism
- Agro Tourism
- Gastronomy Tourism
- Cultural Tourism
- Health Tourism
- Sports Tourism
- Religious Tourism
- Cave Tourism
- Winter Tourism

4.3 ALTERNATIVE TYPES OF TOURISM IN TURKEY

4.3.1 RURAL TOURISM

Nowadays, it is observed that individuals go on trips in order to escape from their everyday lives. However, urbanization, industrialization, harsh business conditions, stress, increasing population, noise and environmental pollution can be listed as reasons why people prefer rural areas in the name of a rest and a peaceful environment (Torun, 2013).

Rural tourism is a concept often confused in the literature. It is possible to come across that rural tourism concept is often associated with or replaced with agricultural tourism, farm tourism, ecotourism, and highland tourism (Doğan and Özaslan, 2017).

In essence, rural tourism refers to activities carried out within the natural environment. Within this environment, a multiplicity of different and intertwined activities can be offered to prospective tourists such as festivals, important events, agricultural events, theatre and different art shows, as well as a wide range of sporting activities (Lane, 1993).

Turkey has significant advantages compared to many other countries in terms of rural tourism given the diversity of its geographical features, and the richness of its historical, cultural, and natural resources. Turkey is able to offer many different tourist products and has a structure that can move out of the sea-sand-sun triangle and provide rural tourism opportunities in many parts of the country. In all regions of the country, there are visually, naturally, and climatically compatible destinations where rural tourism activities can be carried out (Çeken, 2003).

4.3.2 FARM TOURISM

Farm tourism or farm-based tourism activities have been known to occur frequently in recent times. Among the most important reasons for this situation, the favoring of ecological balance throughout the world, the preference of individuals for ecotourism, the protection of nature and the environment, the depletion of resources, and the awareness of local people can be listed (Ahmadova and Akova, 2016).

Farm tourism benefits regional development on the one hand, while farmers gain additional profits. Moreover, by combining agricultural activities and tourism activities, it is ensured that local people and farmers earn extra income along with their own income. Tourists have the chance to live everyday life with farmer families and participate in livestock and some agricultural production activities. On the other hand, recreational activities such as hiking, horse riding, etc. are also offered to tourists (Saygın and Rich, 2018; Cichowska and Klimek, 2011).

Farm tourism strikes as a new type of tourism, which gives importance to the environment along with Agro tourism and rural tourism. There are different definitions related to farm tourism in the related literature. However, as a summary, it can be expressed as a type of tourism in which both farm activities can be carried out and touristic activities are carried out together with the families who own the farm (Ersöz and Bozkurt, 2015).

From the perspective of Turkey, TaTuTa (tarım-turizm-takas/agriculture-tourism-swap) project draws attention. From 2004 to the present, the project is the representative of Worldwide Opportunities on Organic Farms (WWOF) network with more than 11,000 members worldwide. Both with this project and with its own facilities, many farms in Turkey are welcoming guests in an integrated way with tourism (Ahmadova and Akova, 2016).

4.3.3 AGRO TOURISM

Agricultural tourism is a type of tourism seen in the regions where agricultural production activities are carried out. Thanks to agro tourism, agricultural producers have created an additional source of income and the road has been opened for the integration of agricultural production, instead of replacing it with tourism, which is the biggest problem in the rural areas (Civelek and Dalgın, 2013).

Agricultural tourism means visiting agricultural production farms, vineyards, gardens, fields, poultry coops, or local food facilities. In addition, activities such as participating in production activities, traveling, having fun, and even training are carried out during these visits (Türkben et al., 2007).

In another study, it has been argued that agricultural tourism involves activities such as camping, festivals, youth activities, meetings and weddings picnics, local product sales, country walk, bird watching (Yavuzaslanoğlu and Yavuz, 2012). In a different study, it is stated that the tourists who participate in agricultural tourism activities not only participate in agricultural activities or recreational activities, but also have a common interest in the life of the local people for a certain period (Sznajder, et al., 2009).

When examined within the framework of agricultural tourism, Turkey is seen to have an important potential. The geographical conditions and climatic conditions of Turkey, which has a very fertile land for agricultural production also allow tourism activities to be carried out together with agricultural production. Agriculture is carried out in rural areas in many destinations of Turkey. Positive effects such as development, migration prevention, and additional income are noted in these regions and it is observed that agricultural tourism will be increasingly applied (Civelek and Dalgın, 2013).

4.3.4 GASTRONOMY TOURISM

The subject of gastronomy has recently been noted as a highly intriguing concept. There is an increase in activities aimed at integrating gastronomic riches and culinary cultures with tourism. It is stated that experiences on eating and drinking are an important detail in a tourist's travel. Gastronomic tourism brings with it concepts such as chocolate tourism, wine tourism, cheese tourism, and beer tourism. Gastronomic travel to destinations such as

France, Italy, Spain, The Netherlands, and the Far East is observed throughout the world (Şahin and Ünver, 2015).

To make a definition, the whole of the experiences carried out in gastronomic regions with the activities based on eating, drinking, and having fun are expressed as gastronomic tourism. Gastronomic festivals, stands, and markets where local products are exhibited, trips to food and beverage producers are considered products of the relevant type of tourism (UNWTO, 2014).

Turkey is an advantageous country for gastronomic tourism. Due to its culinary culture and historical structure, Turkey is one of the leading countries in gastronomic tourism. Gastronomic travels are carried out to regions famous for their local cuisine such as Adana, Antioch, Sanliurfa, and Izmir (Şahin and Ünver, 2015). However, although Turkey has such a valuable history and culture, the supply and demand for gastronomic tourism is still not at the expected level. The sea-sand-sun trio leaves behind many types of tourism in Turkey (Bucak and Aracı, 2013).

4.3.5 CULTURAL TOURISM

Cultural tourism includes activities such as recognizing the cultural values of a region, acquiring knowledge, making experiences for cultural values, and participating in activities such as sightseeing, publicity (Boz et al., 2017).

Cultural tourism combines many different elements archaeological trips, museum visits, festivals, places with architectural features, artistic activities, religious structures in short, intangible and tangible activities for all cultural heritage are within the framework of cultural tourism (Richards, 2001; Çulha, 2008; Usta, 2012).

Cultural tourism has a place in the literature as a type of tourism that offers tourists the way of life, traditions, and many different elements of different societies while providing participation in cultural activities (Saarinen et al., 2014; Huh, 2002).

In this context, when cultural tourism is examined in Turkey, a wide range of products is observed. Turkey has a profound cultural heritage. The basis of this heritage comes from the civilizations that lived in Anatolia throughout the history. In Anatolia, which catered for life from the dawn of history to the present day, the artifacts of life extending from Göbeklitepe to Rome and Byzantium to the Ottomans have now been transformed into elements which are serving tourism.

4.3.6 HEALTH TOURISM

There are a number of reasons that lead individuals to travel for touristic purposes; among these, health is an increasingly important motivation. Individuals who travel to maintain or improve their health are examined within the scope of health tourism (Özsarı and Karatana, 2013).

Medical tourism is a new subject of research, although people have always traveled to seek medical treatment (Aslanova, 2013). The rapid growth of the tourism sector in the world and in Turkey brings new and different tourism options, and the necessity of adding new types of tourism instead of the ones that are limited to a specific season is emerging. In this context, health tourism is important (Özsari and Karatana, 2013).

It is possible to sort the reasons why individuals with health issue travel abroad as follows (Şahin and Şahin, 2018):

- Quality of treatment,
- Limitations of health insurance,
- Health expenditures,
- The desire to take a holiday combined with high-tech health care treatment,
- Limited opportunities for holidays in climate and geography,
- Thermal facilities and thermal tourism facilities.
- The desire of chronic patients, the elderly, and the disabled to go to an environment that improves the quality of their lives.

4.3.7 SPORTS TOURISM

Sports tourism has been noted as a hugely popular topic in recent years. Different ideas have been developed in terms of defining sports tourism. Sports tourism entered the International Literature in the 1990s (Zgurovski and Akın, 2016).

Sports tourism is expressed as a type of tourism in which there are activities such as sports clubs, trips for the purpose of preseason training (Koca et al., 2007), all kinds of sports activities (Bektas, 2010), national level and international level games, sports-related tours, and championships (Bektas, 2010). In another study, sports tourism is defined as the trips that fans or sportsmen make in order to watch or participate in sporting activities (Gibson, 1998).

Sports tourism attracts a significant number of sports teams, sports organizations, sportsmen, and fans to Turkey and around the world within the concept of holidays and sports. Therefore, World Cups, sports festivals, and many organizations have started to be followed by a significant number of people across the world (Cimilli et al., 2016).

4.3.8 RELIGIOUS TOURISM

Travelling to visit religious centers or to fulfill one's religious duties is expressed as faith tourism (Avcikurt, 1997).

Religious tourism is an important type of alternative tourism. It is often referred together with cultural tourism. It is also involved in cultural tourism in some studies. It is known that religious works or religious places, especially those from the past to the present, are considered cultural heritage because they belong to ancient civilizations. It is therefore also recognized as an integral part of cultural tourism (Richards, 2007).

The geopolitical position of Anatolia and the settlement of many civilizations in Anatolia has resulted in the sacred places of these civilizations to be located within Turkey. The Islamic and Christian cultures and many works belonging to ancient religious structures are located within the borders of Turkey (Büyükşalvarcı et al., 2017).

4.3.9 CAVE TOURISM

Caves are described as natural cavities, usually located underground and within rocks, into which humans can enter. The formation of caves can take thousands of years. It is known that there are quite a significant number of these geographical structures in Turkey (İşçen, 1992).

The number of caves that can be explored and logged throughout the world is around 2000–2500. However, the number of caves in Turkey is estimated to be approximately 40,000. The interest of local and foreign tourists in these caves has started to increase (Kürekçi and Sezen, 2018).

Within the scope of the geographic and temporal diversification of tourism types, caving might hold an important place. There has been an increase in research, especially in recent years. Considering that there are still thousands of caves in Turkey with known and proven health benefits, it is anticipated that caving and cave tourism will be of great interest in the future (Sarı and Demirkaya, 2012).

4.3.10 WINTER TOURISM

Skiing in mountains with a certain height, participation in training activities related to this sport. Visit to regions where winter sports are organized, accommodation, touristic activities in these visits are considered as winter tourism (Ilban and Kaşlı, 2008).

In simpler terms, winter tourism is a type of tourism, which is conducted, especially in the highlands, where snowfall is intense and which consists of mostly sports (Ibragimov, 2001).

Winter tourism means in another words time spent with nature and fresh air. This situation is very attractive for those who spend their daily lives with work life, stress, and in intense city weather. Because the transition from difficult and stressful living conditions to healthy and peaceful winter tourism activities has positive effects in every sense. However, the possibility of participation in winter sports is also noted as a plus value (Mursalov, 2009).

Winter tourism is the most important competitor of sea tourism as of the season. In other words, it is an important type of alternative tourism. From the perspective of Turkey, the world famous winter tourism facilities serve to tourism. Across Turkey, the height and thickness of the snow are above the expected averages and numerous facilities are attracting attention (Tür, 2018).

4.4 CONCLUSION

Although not large enough in the world tourism market, Turkey has significant tourism statistics compared to rival countries. Many destinations host millions of tourists each year. However, a large amount of tourism activities is conducted within the setting of mass tourism (sea-sand-sun). Because of the implementation of mass tourism, the negative economic, social, and environmental effects of tourism are observed. Bed capacities, whose numbers have been increasing year by year remain dormant at times other than the summer months, when tourists arrive intensively. In this context, productivity remains weak. It is inevitable to turn to alternative tourism products so that the capacity in question can be used regularly and efficiently. The most effective solution for combating these problems is to ensure the sustainability of tourism and to diversify tourist products.

In Turkey, which has the capacity to host all kinds of touristic activities with its historical, cultural, and natural richness, it is known that the diversity of tourism in question cannot be adequately utilized and used in tourism industry. The dynamics of the public and private sector need to be mobilized in this regard. Stakeholders in the tourism sector should be encouraged to act together. By applying all the strategies of modern marketing, consumption of different tourist products should be ensured by analyzing and even directing the wishes and requirements of current and potential tourists.

KEYWORDS

- **alternative tourism**
- **sustainable tourism**
- **mass tourism**
- **sea-sand-sun**
- **Turkey**

REFERENCES

Ahmadova, S.; Akova, O. Türkiye'de Organik Ekoturizm Çiftlikleri Üzerine Bir Araştırma. *Karabük Üniversitesi Sosyal Bilimler Enstitüsü Dergisi.* **2016,** *6*(1), 14–29.

Arslaner, E.; Erol, G. Alternatif Turizmin Bazı Türleri Üzerine Bir Değerlendirme. *J. Tour. Gastron. Stud.* **2017,** *5*(4), 422–438.

Aslanova, K. Türkiye'de Sağlık Turizmi Ve Sağlık Turizmi Hukuku. *AVRASYA Uluslararası Araştırmalar Dergisi.* **2013,** *2*(3), 129–145.

Avcıkurt, C. *Turizm Sosyolojisi*; Detay Yayıncılık: Ankara, 2003.

Baykan, E. Alternatif Turizm Kaynaklarından Golf Turizmi ve Kongre Turizmi; Ülkemize Ekonomik Katkıları Açısından Karşılaştırılması: Antalya Örneği; *Yayımlanmış Uzmanlık Tezi. Akdeniz Üniversitesi*: Antalya, 2014.

Bektaş, F. *Kaçkar Havzası Trekking Parkurlarının Spor Turizmi Bakışından Değerlendirilmesi.* Yayımlanmamış Doktora Tezi; Gazi Üniversitesi Eğitim Bilimleri Enstitüsü: Ankara, 2010.

Bramwell, B.; Lane, B. *Rural Tourism and Sustainable Rural Development.* Proceedings from the Second International School of Rural Development. Ireland Channel View Books, London, 1994.
Boz, M.; Yurdakul, S. B.; Turdubekova, S. Yerli Ve Yabancı Turistlerin Sinop'un Kültür Turizmi Çekiciliklerine Ve Altyapısına İlişkin Algıları. *MANAS Sosyal Araştırmalar Dergisi.* **2017,** *6*(5), 49–65.
Bucak, T.; Aracı, Ü. E. Türkiye'de Gastronomi Turizmi Üzerine Genel Bir Değerlendirme. Balıkesir Üniversitesi Sosyal Bilimler Enstitüsü Dergisi. **2013,** *16–30*, 203–216.
Büyükşalvarcı, A.; Şapcılar, M. M.; Altınışık, İ.; Şimşek, Ç. Turizm Öğrencilerinin İnanç Turizmine Yönelik Bakış Açıları. *Selçuk Üniversitesi Sosyal ve Teknik Araştırmalar Dergisi.* **2017,** *14*, 42–61.
Caira, S.; Ferranti, P. Innovation for Sustainable Agriculture and Food Production. *Reference Module in Food Science*; Italy, 2016.
Cichowska, J.; Klimek A. The Role of Agrotourism in the Development and Conversion of Rural Areas. *Polska Akademia Nauk, Oddziaá W Krakowie.* **2011,** *11*, 97–107.
Cimilli, V.; Gülbahçe, Ö.; Şafak, S. Spor Turizmi Açısından Türkiye'nin Kış Sporlarındaki Potansiyeli1in Değerlendirilmesi. *Uluslararası Spor Bilimleri Dergisi.* **2016,** *2*(1), 22–32.
Civelek, M.; Dalgın, T.; Çeken, H. Agro-Turizm ve Sürdürülebilirlik İlişkisi: Muğla Yöresindeki TaTuTa Çiftliklerinde Bir Araştırma. *J. Tour.Gastron. Stud.* **2013,** *1*(3), 23–33.
Cottrell, S. P.; Vaske, J.; Roemer, J. Resident Satisfaction with Sustainable Tourism: The Case of Frankenwald Nature Park, Germany. *Tour. Manag. Perspect.* **2013,** *8*, 42–48.
Çeken, H. *Küreselleşme, Yabancı Sermaye ve Türkiye Turizmi*; Değişim Yayınları: İstanbul, 2003.
Çeken, H.; Karadağ, L.; Dalgın, T. Kırsal Kalkınmada Yeni Bir Yaklaşım Kırsal Turizm ve Türkiye'ye Yönelik Teorik Bir Çalışma. *Artvin Çoruh Üniversitesi Orman Fakültesi Dergisi* **2007,** *7*(1), 1–14.
Çelik, S. Alternatif Turizm. *Uluslararası Sosyal Araştırmalar Dergisi.* **2018,** *11*(56), 193–204.
Çetinel, F. G. Turizm Pazarlamasında Alternatif Yaklaşım Olarak Bölgesel Turizm Pazarlaması. *Anadolu Üniversitesi İktisadi ve İdari Bilimler Fakültesi Dergisi.* **2001,** *17*(1), 398–399.
Çulha, O. Kültür Turizmi Kapsamında Destekleyici Turistik Ürün Olarak Deve Güreşi Festivalleri Üzerine Bir Alan Çalışması. *J. Yasar Univ.* **2008,** *3*(12), 1827–1852.
Doğan, S.; Özaslan, Y. Kırsal Alan Gelişimi Açısından Kırsal Turizm Ve Kırsal Turizmin Dünyadaki Durumu. *Erzincan Üniversitesi Sosyal Bilimler Enstitüsü Dergisi.* **2017,** *4*(Special Edition), 61–78.
Ersöz, S.; Bozkurt, M. Kırsal Turizm Kapsamında Çiftlik Turizmi ve Rize İlinde Uygulanabilirliği. *J. Recreat. Tour. Res.* **2015,** 1–9.
Gibson, H. Active Sport Tourism: Who Participates?. *Leis. Stud.* **1998,** *17*(2–3), 155–170.
Hacıoğlu, N.; Avcıkurt, C. *Turistik Ürün Çeşitlendirmesi;* Nobel Yayınları: Ankara, 2008.
Huh, J. *Tourist Satisfaction With Cultural/Heritage Sites: The Virginia Historic Triangle.* Yayınlanmamış Yüksek Lisans Tezi; Virginia Polytechnic Institute and State University: Virginia, 2002.
İbragimov, M. *Ürün Çeşitlendirmesi Açısından Kış Turizmi ve Almatı Çimbulak-Medeu Örneği.* Yayımlanmamış Yüksek Lisans Tezi. DEÜ Sosyal Bilimler Enstitüsü: İzmir, 2001.
İlban, O. M.; Kaşlı, M. *Kış Turizmi, Turistik Ürün Çeşitlendirmesi;* Nobel Yayın Dağıtım: Ankara, 2008.

İşçen, Y. Mağara Turizmi. *Amfora Aylık Turizm, Tanıtım, Kültür, Çevre Dergisi.* **1992,** *2*, 28.
Karadağ, L. *Sürdürülebilir Turizm Stratejilerinin Geliştirilmesinde Turistik Ürün Çeşitlendirmesinin Önemi: Muğla İli Örneği.* Yayımlanmış Doktora Tezi. Muğla Sıtkı Koçman Üniversitesi, Sosyal Bilimler Enstitüsü, 2016.
Koca, H.; Zaman, S.; Coşkun, O. Erzurum'un Spor-Kamp Turizmi Potansiyeli. *Doğu Coğrafya Dergisi.* **2007,** *12*(18), 205–224.
Kozak, N.; Kozak, M. A.; Kozak, M. *Genel Turizm, İlkeler, Kavramlar;* Turhan Kitabevi: Ankara, 2001.
Külekçi, A. E.; Sezen, I. Bir Ekoturizm Aktivitesi Olarak Mağara Turizmi; Erzurum İli Elmalı. Mağarası. *Süleyman Demirel Üniversitesi Mimarlık Bilimleri ve Uygulama Dergisi.* **2018,** *3*(1), 66–75.
Mursalov, M. *Bir Turistik Ürün Çeşitlendirmesi Olarak Kış Turizmi ve Kış Turizmi Açısından Azerbaycan'ın Guba Turizm Bölgesinin Arz Potansiyeli.* Yayımlanmış Doktora Tezi, DEÜ Sosyal Bilimleri Enstitüsü, İzmir, 2009.
Novas, N.; Gázquez, J. A.; MacLennan, J.; García, R. M.; Fernández-Ros, M.; Agugliaro, F. M. A Real-Time Underground Environment Monitoring System for Sustainable Tourism of Caves. *J. Clean. Prod.* **2017,** *142*(4), 2707–2721.
Özdemir, D. Yeniden Canlandırma Projelerinde Kültür, Turizm ve Emlak Piyasaları Üzerine Kurulu Stratejilerin, Başarı(sız)lıkları Koşullarının İncelenmesi. *Kentsel Dönüşüm Sempozyumu, 11-13 Haziran.* İstanbul Teknik Üniversitesi: İstanbul, 2003.
Özsarı, S. H.; Karatana, Ö. Sağlık Turizmi Açısından Türkiye'nin Durumu. *Kartal Eğitim Ve Araştırma Hastanesi Tıp Dergisi.* **2013,** *24*(2), 136–144.
Öztürk, Y.; Yazıcıoğlu, İ. Gelişmekte Olan Ülkeler İçin Alternatif Turizm Faaliyetleri Üzerine Teorik Bir Çalışma. *Ticaret ve Turizm Eğitim Fakültesi Dergisi.* **2002,** *2*, 183–195.
Richards, G. *Cultural Attractions and European Tourism;* CABI Publishing, New York, 2001.
Richards, G. *Cultural Tourism: Global and Local Perspectives*; The Haworth Press: Binghamton, 2007.
Saarinen, J.; Moswete, N.; Monare, M. Cultural Tourism: New Opportunities for Diversifying the Tourism Industry in Botswana. *Bull. Geogr. Socio Econ. Ser.* **2014,** *26*, 7–18.
Sarı, C.; Demirkaya, H. Mağara Turizmi ve Antalya Mağaralarının Turizm Potansiyeli. *II. Disiplinlerarası Turizm Araştırmaları Kongresi.* 12–15 Nisan; Antalya, 2012.
Savgın, C.; Zengin, B. Türkiye'de Çiftlik Turizmine Yönelik Kavramsal Bir Model Önerisi. *Akademik Sosyal Araştırmalar Dergisi* **2018,** *6*(86), 239–258.
Soykan, F. Avrupa'da Kırsal Turizme Bakış Kazanılan Deneyim. *II. Balıkesir Ulusal Turizm Kongresi,* **2006,** 71–87.
Sznajder, M.; Przezborska, L.; Scrimgeour, F. *Agritourism*; Ama Data Set LTD. UK, 2009.
Şahin, G. G.; Ünver, G. Destinasyon Pazarlama Aracı Olarak "Gastronomi Turizmi": İstanbul'un Gastronomi Turizmi Potansiyeli Üzerine Bir Araştırma. *J. Tour. Gastron. Stud.* **2015,** *3*(2), 63–73.
Şahin, Ö. U.; Şahin, M. Türkiye'de Sağlık Turizminin Potansiyeli Ve Geleceği: Swot Analizi. *J. Aware.* **2018,** *3*(Special Edition), 287–300.
Torun, E. Kırsal Turizmin Bölge İnsanına Katkıları. *KMÜ Sosyal ve Ekonomik Araştırmalar Dergisi* **2013,** *15*(24), 31–37.
Tür, E. Türkiye Ve İsviçre'nin Kış Turizmi Faaliyetleri Açısından Değerlendirilmesine Yönelik Bir Çalışma. *Uluslararası Global Turizm Araştırmaları Dergisi.* **2018,** *2*(1), 10–31.

Türkben, C.; Gül, F.; Uzar, Y. Türkiye'de Bağcılığın Tarım Turizmi (Agro-Turizm) İçinde Yeri ve Önemi. *KMÜ Sosyal ve Ekonomik Araştırmalar Dergisi.* **2007,** *14*(23), 47–50.

Ulusan, Y; Batman, O. Alternatif Turizm Çeşitlerinin Konya Turizmine Etkisi Üzerine Bir Araştırma. *Selçuk Üniversitesi Sosyal Bilimler Enstitüsü Dergisi.* **2010,** *23*, 243–260.

Usta, N. İnanç Turizminden Kültür Turizmine: Midilli Adası'nın Agiasos Dağ Köyü Örneği. *Karamanoğlu Mehmetbey Üniversitesi Sosyal ve Ekonomik Araştırmalar Dergisi.* **2012,** *22*, 163–168.

Yalçın, A. Sürdürülebilir Kalkınma İçin Yeşil Ekonomi Düşüncesi Ve Mali Politikalar. *Çankırı Karatekin Üniversitesi İktisadi ve Idari Bilimler Fakültesi Dergisi.* **2016,** *6*, 1–27.

Yavuzaslanoğlu, E.; Yavuz, M. Karaman'ın Tarım Turizmi Yönünden Potansiyeli, *KMÜ Sosyal ve Ekonomik Araştırmalar Dergisi.* **2012,** *14*(22), 31–33.

You, H.; Zhang, X. Sustainable Livelihoods and Rural Sustainability in China: Ecologically Secure, Economically Efficient or Socially Equitable?. *Resour. Conserv. Recycl.* **2017,** *120*, 1–13.

Zgurovski, K.; Akın, A. Spor Turizminin Gelişimine Yönelik Alan Araştırması (Gaziantep Örneği). *Seyahat ve Otel İşletmeciliği Dergisi.* **2016,** *13*(1), 47–59.

UNWTO. *Report, World tourism Barometers*; 2014 (accessed July 5, 2019).

yigm.ktb.gov.tr/TR-9851/turizm-istatistikleri.html (accessed Aug 16, 2019).

CHAPTER 5

AN EVALUATION OF CITTASLOW (SLOW CITY) MOVEMENT IN TURKEY IN TERMS OF SUSTAINABLE DEVELOPMENT

A. KARATAŞ[1] and G. GÜNEY[2*]

[1]*Marketing and Advertising Department, Muğla Sıtkı Koçman University, Muğla, Turkey*

[2]*Foreign Trade Department, Muğla Sıtkı Koçman University, Muğla, Turkey*

[*]*Corresponding author. E-mail: gulayormeciguney@gmail.com*

ABSTRACT

There is a mutual relationship between Cittaslow's philosophy and sustainable development, which considers a balance of production and consumption that enables future generations to meet their needs based on consuming without destroying. Turkey is a developing country. It requires an understanding of sustainable development and socio-economic development. Moreover, it has many cities that deserve to be candidates of Cittaslow with their rich histories, socio-cultural richness, and local values. This paper will focus on the potential role of this new era on sustainable tourism in Turkey.

"Speed anesthetizes. We are everywhere and nowhere." (Sayar, 2018:37).

5.1 INTRODUCTION

"We cannot disturb nature. We cannot rush rain, wind, and day. Everything is realized in its own time. Similarly, if we stick to our world's natural rhythm and respect them, we will learn to live more meaningfully and differently" (Sayar, 2018:52).

Today all of the objects, space, and phenomena, which meet our needs, have brought new perspectives. The Cittaslow movement started in 1999 in Orvieto and adopted a philosophy of slow movement, which aims to preserve the quality of each city's sense of identity, making life easy, fun, and sustainable for the city-dwellers given that cities are places shaped by their natural, cultural, and historical values, as well as their self-identities. However, as a result of the consumer patterns imposed with globalization, cities are faced with the danger of alienation of their self-identities and past. The Cittaslow movement, which grounds on and brings the city's self-identity to the forefront, loyal to the principles of conservation, and utilization based on the philosophy of sustainability proposes a feasible alternative to protect the cities against this danger. Slow Food was the inspiration for the Cittaslow movement that emerged in Italy. Cittaslow is the adapted version of the Slow Food movement which emerged as a criticism of fast-paced life and consumption patterns, and developed as a resistance to urbanization processes. The slow philosophy, which constitutes the foundation of the Cittaslow, is based on the principles of conservation and utilization and its goal is to transfer the city's unique and distinctive features and resources to the next generations. In this context, there is a mutual relationship between Cittaslow's philosophy and sustainable development, which considers a balance of production and consumption that enables future generations to meet their needs based on consuming without destroying.

Turkey is privileged to have 17 of its cities among the 252 Cittaslow cities worldwide. Turkey has hosted many civilizations for thousands of years making its cities rich in historical, socio-cultural values, and natural resources, and thus matching the Cittaslow criteria.

5.2 WHAT IS THE SLOW CITY MOVEMENT?

The Cittaslow movement was started in 1999 in Orvieto, Italy. The name is based on a combination of Italian "citta" meaning city or town and "slow" in English (Pink and Servon, 2015). The logo of slow city is a "snail." The Cittaslow movement and philosophy, The Slow City Association (The Cittaslow Union) and membership criteria, and the organizational pattern went beyond the borders of Italy in a short time and spread mainly to Europe and all over the world.

The foundation of the Cittaslow movement is based on the "Slow Movement," it is therefore important to examine the content of the slow movement to better comprehend the Cittaslow concept. The slow movement is a critical response to the fast-paced lifestyle of a consumerist society imposed on societies by globalization. Carlo Petrini called this phenomenon "virtuous globalization." As an activist, Carlo Petrini not only inspired the emergence of the Slow Food and the Cittaslow movements but he also led them (Jones et al., 2003). Thus, the vital and urban philosophy of the Cittaslow is not to do everything slowly or calmly but to realize activities in different areas of life more satisfactorily, more in-depth, and at a more appropriate speed (Baldemir et al., 2013). In other words, the slow movement does not act against globalization strictly and destructively but rather pursues the goal of promoting the adoption of consumption patterns based on the principles of conservation and utilization from a social and economic perspective.

Cittaslows are usually lesser known and constitute ordinary cities. These cities seek to provide easy and pleasant lives for its residents and to raise awareness to protect their sense of self-identity. Thus, the common themes that Cittaslows in different countries share are the protection of their self-identities, sustainable development, high quality of life, and local development (Knox, 2005). The "slow" concept as part of the Cittaslow philosophy involves an understanding of protecting the traditional way of life while embracing the positive aspects of modern life and rejecting immobility or backwardness (Honore, 2008). It is an approach that contributes to making the world a better place to live.

The Slow Food movement started in 1986 preceded the Cittaslow movement. As globalization enters a struggle for dominance over the local, this process has also exposed the vulnerability of self-sufficient communities which run the risk of losing their ties and their self-identities (Radstrom, 2011). In order to minimize this risk and preserve their self-identities, the main goal is to protect and sustain what is already established. One of the most basic cultural elements constituting self-identity is local dishes. In this sense, the Slow Food movement aims to discover those foods at risk of disappearing and perish with the advance of consumer society created by globalization and sustaining them by making them economically viable within the principles of conservation and utilization (Honore, 2008). The Slow Food movement has direct links with more than 160 countries in the world and has approximately 1500 local chapters called "convivium" and more than 100,000 members in more than 120 countries in the world (Radstrom, 2011). The Slow Food movement is one of the first and most organized components of the slow movement.

The main philosophy of the Slow Food movement is to protect the local's traditional produce and promote local producers. This understanding was interpreted within the framework of urban dynamics and inspired the emergence of the Cittaslow movement. Slow Food and Cittaslow movements are formations that are based on similar philosophies and in this sense they have common themes and attitudes. However, while the Slow Food movement is a civil society movement, Cittaslow is a local administration approach based on a similar philosophy; thus, it is an institutional structuring which includes public enterprises (Radstrom, 2011). In this regard, Cittaslow will exist as long as it is supported within the local administrations and managed correctly through coordinated actions.

Cittaslow is based on the principles of preserving and sustaining cities that are unique by using their local dynamics, historical texture, past, particular values, and potentials and having distinctive and unique characteristics rather than homogenous cities that become the same as the globalization process.

The Cittaslow movement supports the incorporation of abandoned or endangered natural, historical, and cultural values into the world economy by associating them with sustainable development and conservation policies (Mayer and Knox, 2006). There are mutually supporting goals between Cittaslow and sustainable development which adopt sustainability in terms of its social, economic, and environmental aspects.

5.3 WHAT ARE THE MEMBERSHIP PROCESS AND REQUIREMENTS FOR CITTASLOW AND THE MEMBERSHIP CRITERIA?

The Cittaslow movement aims to minimize the impact of homogenization that has occurred with globalization and which threatens local values (Nilsson et al., 2011). The Cittaslow movement is the driving force of the sustainable development for a city because within its philosophical framework, this movement also serves the goals of sustainable development by conserving local values, historical texture, socio-cultural structure, biological diversity, local food, and the environment.

In 1997, at the World Slow Food Congress in Orvieto, Carlo Petrini put forward the idea of bringing together cities that adopted the philosophy of slowness and calmness under a common roof. The first concrete step was taken with Carlo Petrini's initiative to establish an association of slow cities (Miele, 2008). The Slow City Association, a union of municipalities was first proposed by Carlo Petrini the founder of Slow Food, who along with Paolo Saturnini, the mayor of Greve in Chianti, Italy, and the mayors of "Bra, Greve in Chianti, Orvieto, Positano" laid the foundations of the Slow City Association. As a result of this effort, they became pioneers in the emergence of the Cittaslow movement. As the Cittaslow movement was born in Italy, it enabled the organization to expand more in this country, and thereafter its principles and philosophy were spread throughout Europe and all over the world.

There are two requirements to meet in order to become a member of the Slow City Association. The first one being that the town must have a population under 50,000 and the second one, that it must not be a city center. After meeting these requirements, a set of criteria for certification are identified and within the framework of compliance of standards, the city obtains the Cittaslow title/membership through an application, supervision, and approval process.

With the intention of improving the quality of life, the Cittaslow movement put forward the idea of cities evaluating themselves and presenting an alternative development model which could be applied on a national scale. Today, the Cittaslow movement has become a network in 30 countries with 252 members (www.cittaslowturkiye.org), which provides an alternative to a consumerist society imposed by globalization by applying the Slow Food philosophy on an urban dimension and thus spreading worldwide as an organized structure.

Nowadays, the Cittaslow movement is an international municipal association in which cities with a population under 50,000 can join. In order to become a member of the association, the cities are required to develop and apply their projects in accordance with the membership criteria. The activities carried out by the cities within the framework of the criteria are scored and the city has to obtain at least 50 to become a member. The criteria identified in 1999 by the association became more universal as a result of their expansion worldwide instead of just Europe and Italy. After the study carried out by an International Scientific Committee was approved by the association, the updated criteria went through a trial period for one year during the first stage (www.cittaslowturkiye.org).

As stated before, in order to become a member of the Cittaslow Association, the city must meet the specific criteria. These criteria consist of 72 requirements subdivided into seven main areas with three prerequisites.

These criteria can be compiled under seven titles:

- Environmental Policy
- Infrastructure Policy
- Policies for Urban Quality
- Policies for Agriculture, Tourism, Artisan, and Handicrafts
- Policies for Hospitality, Awareness, and Training
- Policies for Social Cohesion
- Partnerships

Moreover, three prerequisites have to be accomplished. These three prerequisites include the following: meeting the requirements for the slow city action-identity campaigns, supporting and organizing the Slow Food movement, using the slow city logo during the official correspondences between the member cities, and including a section which explains the slow city philosophy on their Internet sites (Keskin, 2012:93).

There are three main organs in the organizational structure of the association: The International Coordinating Committee that also acts as a decision-maker and is made up of representatives of national networks; the International Scientific Committee that composes the scientific substructure and draws the outline of the movement, and the General Secretary (Headquarters) that is responsible for the association's operations. The Cittaslow association is composed of national networks. If there are three Cittaslows in a country, a national network can be established. The national networks administer the candidacy process in their own country and establish communication between the Cittaslows in their country and the Headquarters. They can carry out projects with the members to expand and develop the Cittaslow movement in their country (www.cittaslowturkiye.org). All criteria and projects aim to sustain the local identity and comply with the global one. The Cittaslow movement takes place as an alternative development model as it serves the purpose of sustainability within the framework of conservation and utilization from local to global. In other words, there is a tight and nurturing conceptual bond between the Cittaslow movement and sustainable development.

5.4 WHAT IS THE RELATIONSHIP BETWEEN SUSTAINABLE DEVELOPMENT AND THE CITTASLOW MOVEMENT?

Among the most important and known definitions, sustainability means "to be engaged in an activity to maintain continuity of the existing resource" (Meadowcroft, 1997). Moreover, sustainability can be defined as having the right to use resources but not the right to consume them because the greatest threat posed by global consumption habits to humanity is an understanding of production and consumption processes, which results in the destruction of resources. The balance between humankind and nature is deteriorating at a rapid pace, threatening future generations. There is an increasing need for development models that meet the needs of future generations. In other words, it has become obligatory to create sustainable development models for humanity. The concept of "sustainable development" is defined as "a development that meets the needs of the present without compromising the ability of future generations to meet their own needs" (WCED, 1987). According to the definition of the United Nations' Commission on Environment and Development from 1987, "humanity, without endangering the ability to respond to the requirements of future generations, can have the ability to sustain development by meeting their daily needs." Social structure, economy, and environment are three basic components of the sustainable development approach. The sustainable development approach can only be realized through the integration of all economic and social policies of a country with the environmental policies and strategies.

Economic development and ecosystems are regarded as two nurturing sources. Naturally, sustainable development is only possible with a sustainable ecosystem. In this sense, the parallelism between the sustainability principles and the Cittaslow movement is emphasized and thus, we can maintain that the Cittaslow movement has emerged as an alternative model for sustainable development.

According to Hardi and Zdan (1997), sustainable development is expressed in three categories: economic, environmental, and social effects/dimensions. This triple categorization provides an opportunity for better analysis and understanding of the concept.

The concept of sustainable development was first brought to the agenda by the World Commission on Environment and Development in 1987 with the report Common Future, also known as the Brundtland Report, and has been debated since and gained importance.

Cities which are growing uncontrollably particularly threaten the future of underdeveloped and developing countries. On the other hand, the mega-cities of the developed countries become the source of important problems such as expansion, pollution, transportation, and social segregation.

The only solution to this chaotic situation is to align sustainable development policies with urbanization (Yazar, 2009:115). Cittaslow is a movement, which has the ability to develop solutions to this problem.

The unsustainable production and consumption patterns generated by globalization expose many negative impacts; in particular, the consumption habits imposed by a consumerist society based on rapid consumption have reached levels that are threatening future generations with the damage caused to the environment.

In addition, local and authentic identities have started to erode with the phenomena of global consumption habits, homogenization, and the dominance of global brands. Along with the risk of extinction of local identities, globalization poses a social and cultural threat.

The Cittaslow movement is an alternative development model for cities that do not want to transform themselves into homogeneous spaces, but develop without losing their local identity. In line with its philosophy and goals, the Cittaslow movement promises development by providing these cities with the opportunity to utilize environmental quality via conservation, creating income and employment within the framework of local values, and preserving historical and cultural texture. Moreover, it can be stated that when the Cittaslow movement's principles and plans are examined, it provides many of the "Indicators of Sustainable Development" cited by the UN, OECD, and the EU, and are even some superior in some aspects. In this sense, when the Cittaslow movement is analyzed, it can be established as the motor of sustainable development.

In order to maintain sustainability, the relationship between the processes related to the economic, social, and ecological dimensions should be evaluated as a whole; and the objectives and targets should be in harmony with these dimensions during the decision-making and policy-making activities (Keskin, 2012:84). Urbanization and local government activities are very important to directly achieve the goals and targets of sustainability.

The slow city model was presented as a local development model to create sustainable cities and lifestyles. They acknowledge that the creation of environments can make individuals content, be more productive, and blend in with community life (Ekşi and İbrahim, 2019). However, unexpected or even undesirable situations may occur and fail to fulfill the objectives of this model.

That is to say, sometimes contrary to expectations, the concept of slow city can have negative effects on sustainable development. For example, Şahin and Kutlu's study (2014) determined that the positive relationship between the concepts of sustainable development and of Cittaslow was disrupted due to two factors. They stated that in order to become a member of the Cittaslow Association of Cities, the requirement that the city must have a population of less than 50,000 (without exception) negatively affected the feasibility of Cittaslow in terms of large cities. Moreover, they concluded in their study that Cittaslow-based sustainable development does not extend to big cities with significant physical, environmental, social, and economic problems.

In another study, it was concluded that "if the unique and local characteristics of a city that make it a Cittaslow are managed without considering the conservation and utilization balance, that is being away from the understanding of sustainable tourism, unlike the Cittaslow movement, a locomotive for sustainable development, it can turn into a movement that will interrupt sustainable development" (Karataş and Günay, 2018).

When evaluated from the perspective of sustainable development, it can be argued that it is only possible for a city to develop by protecting its own identity. The continuity of the production process and the revival of social and economic life are important to keep alive and sustain cities because cities without economic power are no longer liveable. Due to unemployment and a lack of social opportunities in small cities, the fact that younger generations preferred/migrated to big cities accelerated this process.

5.5 HOW DID THE CITTASLOW MOVEMENT DEVELOP IN TURKEY?

The Cittaslow model presented sustainable cities and lifestyle as a local development model promotes the creation of environments that can make individuals content as well as those around, be productive, and integrated with community life (Ekşi and İbrahim, 2019). It is required that all movements, which defend the philosophy of slowness, must act together through the application of the Cittaslow criteria in all cities with an awareness of sustainability (Doğutürk, 2010). The Cittaslow movement in Turkey started in Seferihisar on November 28, 2009 when the town was certified as a Cittaslow (Ak, 2017).

Tunç Soyer, the former mayor of Seferihisar, implemented a model town plan during his first year of mayorship (2009), bringing the Cittaslow movement, an international local development model and enabling it to expand to Turkey. Soyer stated that Seferihisar was chosen to be the first Cittaslow in Turkey in 2009 and thus they had an opportunity to become the capital city of Turkey. He said, "The Cittaslow philosophy considers the nature and historical characteristics of the city as its greatest wealth and is based on the preservation of the city's unique identity. It manifests that try to resemble nowhere, bring your own features to the forefront, become a brand, and create awareness. We started to receive many applications. We called Turkey to action and said, 'Turkey could become the paradise of Slow Cities'. After long hours of work, we have expanded to all regions and Northern Cyprus. We are still receiving many applications. We will continue to introduce small and beautiful cities of Turkey to Cittaslow brand and the world." (http://t24.com.tr).

The first "Cittaslow's mayorship in Turkey carried Soyer" to the Vice Presidency of the Cittaslow Association in Italy in 2013 enabling him to lead the Cittaslow movement in Turkey.

A total of 17 towns in Turkey have been accepted as Cittaslow since March, 2019: Muğla Akyaka, Isparta Eğirdir, Çanakkale Gökçeada, Sinop Gerze, Bolu Göynük, Şanlıurfa Halfeti, Bolu Mudurnu, Ordu Perşembe, Artvin Şavşat, İzmir Seferihisar, Sakarya Taraklı, Erzurum Uzundere, Kırklareli Vize, Isparta Yalvaç, Aydın Yenipazar, Muğla Köyceğiz, and Bitlis Ahlat.

It is inevitable that the approach established in the Cittaslow concept "supports the idea of development of the cities that act against banalization caused by globalization by protecting their own identities" and "unearths the cities' characteristics that distinguish them from millions of other cities in the world by preserving their local cuisine, tradesmen, traditions, customs, and history" will make positive social and economic contributions to Turkey (www.hendesedergisi.com).

The locality of a city is an indicator for the region to use its production capacity and obtain economic value from it because the Cittaslow philosophy involves being economically self-sufficient. The economic activities of slow cities in terms of economic production in Turkey can be described as follows:

TABLE 5.1 Cittaslows in Turkey.

Cittaslows in Turkey	Acceptance date	Web addresses
İzmir Seferihisar	2009	http://seferihisar.bel.tr/
Muğla Akyaka	2011	http://www.akyaka.bel.tr
Çanakkale Gökçeada	2011	http://www.gokceada.bel.tr
Sakarya Taraklı	2011	http://www.tarakli.bel.tr
Aydın Yenipazar	2011	http://www.yenipazar.gov.tr
Isparta Yalvaç	2012	http://www.yalvac.bel.tr
Ordu Perşembe	2012	http://www.persembe.bel.tr
Kırklareli Vize	2012	http://www.vize.bel.tr
Şanlıurfa Halfeti	2013	http://www.halfeti.bel.tr
Artvin Şavşat	2015	http://savsat.bel.tr
Erzurum Uzundere	2016	http://www.uzundere.bel.tr
Isparta Eğirdir	2017	http://www.egirdir.bel.tr
Bolu Göynük	2017	http://goynuk.bel.tr
Sinop Gerze	2017	http://gerze.bel.tr
Bolu Mudurnu	2018	http://mudurnu.bel.tr
Bitlis Ahlat	2019	http://ahlat.bel.tr
Muğla Köyceğiz	2019	http://koycegiz.bel.tr/

Source: Adapted from Burkut (2017).

1. **İzmir Seferihisar:** While the basis of economic activities is agriculture and olive cultivation, citrus and artichoke cultivation, greenhouse production including mainly ornamental plants and stockbreeding have become an important source of income in recent years.

   Moreover, fishing and tourism have become one of the most important sectors contributing to the district economy (www.seferihisaradair.com). In addition, Seferihisar also has a branded tangerine, rich resources in terms of solar, geothermal and wind energy, as well as historical richness (https://cittaslowturkiye.org/cittaslow-seferihisar/).
2. **Muğla Akyaka:** Agriculture and tourism are among the most important economic activities. Oranges, lemons, olives, sesame seeds, and fishing are among the most important sources of income. It is a place that respects nature, preserves its architectural

structure, has strong infrastructure services, and focuses on sustainable and alternative tourism (www.cittaslowturkiye.org).

3. **Çanakkale Gökçeada:** The majority of the island's population consists of workers, civil servants, as well as people engaged in agriculture, stock breeding, and boarding-houses, which have developed greatly in recent years. The lack of large industrial enterprises in Gökçeada has led the islanders to alternative sources of income. The main sources of income and production areas are olive cultivating, beekeeping, viticulture, organic farming, and fishing (www.e-gokceada.com).
4. **Sakarya Taraklı:** Due to the lack of a young population, Taraklı does not have much diversity in terms of economic activities. Agriculture and stockbreeding are the main sources of livelihood. "The Taraklı linen," which was repopularized through the Public Education Centre, plays a role in generating income for the women. In addition to thermal tourism, plateau and cave tourism as well as paragliding on the Karagöl Plateau are among the other tourism resources of the district (www.hendesedergisi.com).
5. **Aydın Yenipazar:** Its economy is largely based on agriculture and stockbreeding. Production of cotton, olives, figs, tomatoes, peppers, citrus fruits, and pistachios grown in the district makes important contributions to the economy of the district (www.yenipazar.bel.tr).
6. **Isparta Yalvaç:** The district's income sources are agriculture, stockbreeding, aquaculture, and handicrafts (leather trade). As irrigated agriculture, fruit, vegetables, sugar beet, alfalfa, sainfoin, and corn are grown. Dry farming involves cereals and legumes. Due to the fact that a part of Lake Eğridir is located within the boundaries of Yalvaç district, fishing has been included as a source of income in the region (www.wowturkey.com).
7. **Ordu Perşembe:** The district's economy is based on agriculture. The most important crops in agriculture are hazelnut, kiwi and strawberry, which have gained importance in recent years. In addition to the abundance and variety of agricultural products, Perşembe is known as the Fishing Centre. Moreover, it is a trade center. (www.cittaslowturkiye.org).
8. **Kırklareli Vize:** Two-thirds of its townspeople, who live in the countryside, work on agriculture and forestry. The main crops are wheat, sugar beet, sunflower, and barley. Vegetables and fruits

are also produced. Livestock occupies an important place in the district. Small-scale fishing takes place on the coast. Large coppice oak forests make a significant contribution to the district's economy (www.trakyanet.com).

9. **Şanlıurfa Halfeti:** Industry has not developed in the district. There are no facilities other than small-scale flour mills and briquette plants. The district economy is generally based on agriculture and stockbreeding. Dry farming is done on 98% of the total arable land and 2% of the land is shared for irrigated agriculture. The main products of dry agriculture are wheat, barley, lentils, pistachios, vineyards, and olives (www.urfafx.mekan360.com).
10. **Artvin Şavşat:** The meadow and pasture areas in the district made livestock farming the most important source of income. The rich flora of the region creates suitable environmental conditions for beekeeping. Forests and forestry activities directly or indirectly affect all villages (www.cittaslowturkiye.org).
11. **Erzurum Uzundere:** Many different types of fruit are grown in the district. Agriculture and stockbreeding is an income source for local people. Uzundere has the best features of nature, history, sports and culture with its geographical location, nature, traditional rural life culture that has existed for centuries, natural sports, and historical artifacts (www.cittaslowturkiye.org).
12. **Isparta Eğirdir:** Egirdir has a strong local economy. The most important source of income is the export of apple and fishery products. In addition, other important economic activities are stockbreeding, crafts, and forest products. While most people in the countryside are engaged in agriculture, the majority of them weave carpets in their remaining time. Egirdir is one of the leading tourism centers of Isparta Province, including the entire Lake District. Tourism occupies an important place in the district economy (www.ispartakulturturizm.gov.tr).
13. **Bolu Göynük:** Although the district is not developed economically due to its geographical structure and transportation difficulties, 80% of the district's population makes a living from agriculture and animal husbandry and 20% from the forestry sector (www.goynukkaymakamligi.gov.tr).
14. **Sinop Gerze:** Agriculture, stockbreeding, fishing, textile industry, and tourism are at the forefront of the district economy. Most of the

district's land is forested and arable land. The presence of natural beaches in Gerze increases the marine tourism potential of the district. Kozfındık and Bozluk Highlands, as well as mountain areas in the district are significant contributors to the development of tourism of plateau and mountain sports (www.cittaslowgerze.com).

15. **Bolu Mudurnu:** Mudurnu, which developed as a commercial center and military area at the junction of the historical Silk Road and other important roads, advanced as an important Turkish-Islamic guild (or Ahi order)-artisan city and cultural center of the Early Ottoman Period. Today, visitors prefer it due to its faith (religious) tourism and hot springs (www.mudurnu.bel.tr/).
16. **Bitlis Ahlat:** The economic structure of Ahlat is generally based on agriculture and stockbreeding. The western part of the settlement is completely covered with agricultural fields. Although livestock is mostly raised in villages, there are many who deal with livestock in Ahlat. Especially those families who migrated from villages continue their activities in Ahlat. In the central districts, those who are engaged in trade and those working in government agencies are becoming more relevant (www.ahlat.bel.tr).
17. **Muğla Köyceğiz:** The biggest source of income of the district is agriculture. 85% of the district's population lives in the neighborhood and earns a living through agriculture, animal husbandry, forestry, migratory beekeeping, mullet and trout fish production, citrus fruits, and tourism. Polyculture farming is carried out in the district and the climate and geographical structure are suitable for growing many products (http://koycegiz.bel.tr).

Before, the slow city concept in Turkey was considered to be located only in the western region. However, this assessment has started to lose its meaning with the increasing number of slow cities from north to south and west to east.

Among the 11 Vice Presidents of the International Coordination Committee of Cittaslow Association, one of the vice presidents is from Turkey, which is highly significant. The criteria required to be a Cittaslow such as environmental policies, infrastructure, the quality of texture of the city, supporting local products, and hospitality can increase awareness to maintain their values in Turkey (social, cultural, and economic) (www.ntv.com.tr). Turkey has many cities, which stand out due to their historical richness, local products, cultural values, having a potential to be self-sufficient, and possessing a unique self-identity. In this sense, as stated previously, Turkey can be considered a Cittaslow paradise due to the high potential of applicability. In addition, due to being a developing country and its need for sustainable development, the alternative model presented by the Cittaslow promises a bright future for Turkey.

5.6 CONCLUSION

> *"A new ego that sets the rhythms of life to the rhythms of the market but attaches importance to the values that can be bonded to money is expanding in the world with a global wind" (Sayar, 2018:38).*

Today, globalization changes cities and exposes them to the risk of losing their local values. Global rapid consumption not only erodes these values but also leads to the consumption of resources that cannot be readily replaced. In addition, the rapid changes that have occurred with fast-paced lifestyles create the phenomenon of disidentification in cities. In the global system, cities' unique characteristics disappear and are replaced by homogeneity among cities.

The Cittaslow movement, which emerged in Italy as a response to the negative impacts of globalization on human life such as rapid consumption, loss of resources, erosion of values, erosion of local culture, alienation, and disidentification can be expressed as an alternative solution model.

The prerequisite for Cittaslow is the sustainability of local identities, conservation of resources and the management of cities with the principles of conservation and utilization that serve the concept of sustainability. In this sense, Cittaslow is also a meaningful concept for sustainable development where social, economic, and environmental resources are positioned considering the production and consumption balance in a way that allows the use of resources for future generations. It is considered that there is a nurturing relationship between Cittaslow and sustainable development because they serve similar goals and common principles.

Turkey is a developing country. It requires an understanding of sustainable development and socio-economic development. Moreover, it has many cities that deserve to be candidates of Cittaslow with their rich histories, socio-cultural richness, and local values. As stated before, out of 272 Cittaslows in the world, 17 are in Turkey. In terms of preserving a city's self-identity and providing an ability to be self-sufficient, Cittaslow becomes a significant model to achieve sustainable development goals considering Turkey's potential.

However, these cities must be administered by taking into account the principles of conservation and utilization so that the Cittaslow movement can serve the goals of sustainable development. Otherwise, it becomes a phenomenon that is criticized as simply creating a new tourism attraction. It can encounter results that are presented to the global market and will lead to space consumption.

A contribution can be made if the relationship between the concept of sustainable development discussed above, with its social, economic, and environmental aspects and the growth of slow cities, is managed correctly. Another important all and only prerequisite is to prevent misunderstanding of the playing rules in the market system.

KEYWORDS

- **Cittaslow (slow city)**
- **sustainable development**
- **economic development**
- **Turkish Tourism**

REFERENCES

Ak, D. Uluslararası Sosyal Araştırmalar Dergisi. *J. Int. Soc. Res.* **2017,** *10* (52), 884–903.

Baldemir, E.; Kaşmer Şahin, T.; Kaya, F. Yavaş Şehir Olma Durumunun Analitik Hiyerarşi Süreci ile Değerlendirilmesi. *Ekonomi ve Yönetim Araştırmaları Dergisi* **2013,**. *2* (1), 29–50.

Burkut, E. B. (2017), Assessment of Impact Cultural Heritage and Tourism in Slow City (Cittaslow): The Case of Turkey, ICCHT 2017 Internatıonal Congress on Cultural Herıtage and Tourısm Konya: Turkey, May 19–21, 2017.

Cittaslow International Association (2018). http://www.Cittaslow.Org/Section/Association (accessed Dec 24, 2018).

Ekşi, Y. E.; İbrahim, B. A. Z. Cıttaslow-Sakin Şehirlerde Yaşanabilirlik Üzerine Bir Araştırma. Teknoloji ve Uygulamalı Bilimler Dergisi **2019,** *01* (02), 71–76.

Hardi, P.; Zdan, T. *Assessing Sustainable Development*; International Institute for Sustainable Development: Winnipeg, 1997.

Honore, C. *YAVAŞ! Hız Çılgınlığına Başkaldıran Yavaşlık Hareketi*, (çev: Esen Gür); Alfa Yayınları: İstanbul, 2008.Jones, P.; Shears, P.; Hillier; et al. Return to Traditional Values? A Case Study of Slow Food. *Br. Food J.* **2003,** *105* (4/5), 297–304.
Keskin, E. B. Sürdürülebilir Kent Kavramına Farklı Bir Bakış: Yavaş Şehirler. *Paradoks Dergisi* **2012,** *8* (1), 81–99.
Knox, P. L. Creating Ordinary Places: Slow Cities in a Fast World. *J. Urban Des.* **2005,** *10* (1), 1–11.
Mayer, H.; Knox, P. L. Slow Cities: Sustainable Places in a Fast World. *J. Urban Affairs* **2006,** *28* (4), 321–334.
Miele, M. Cittaslow: Producing Slowness against the Fast Life. *Space Polity* **2008,** *12* (1), 135–156.
Nilsson, J. H.; Svärd, A.; Widarsson, Å.; et al. Cittáslow' Eco-Gastronomic Heritage as a Tool for Destination Development. *Curr. Issues Tourism* **2011,** *14* (4), 373–386.
Ökmen, M. Küreselleşme Sürecinde Yerelleşme Eğilimleri ve Yerel Yönetimler. *Yerel Yönetimler Üzerine Güncel Yazılar I: Reform, (24)*; Özgür. H., Kösecik, M. Eds; 2005; 539–564.
Pink, S.; Servon, L. J. Cittaslow: Going Glocal in Spain. *J. Urban Affairs* **2015,** *37* (3), 327–340.
Radstrom, S. A Place-Sustaining Framework for Local Urban Identity: An Introduction and History of Cittaslow. *Ital. J. Plann. Pract.* **2011,** *1* (1), 90–113.
Şahin, İ.; Kutlu, S. Z. Cittaslow: Sürdürülebilir Kalkınma Ekseninde Bir Değerlendirme. *J. Tourism Gastronomy Stud.* **2014,** *2* (1), 55–63.
Sayar, K. *Yavaşla, Bu Dünyadan Bir Defa Geçeceksin*, (Genişletilmiş Baskı) Kapı Yayınları 589, Psikoloji 12, 2018.
www.cittaslowturkiye.org. (accessed May 12, 2019).
WCED *Dünya Çevre ve Kalkınma Komisyonu: World Commission on Environment and Development*; Our Common Future, Brundtland Report, Oxford & New York: Oxford University Press, 1987.
Yazar, K. H. Sürdürülebilir Kent: Göstergeleri, Formu Ve Planlama Süreci, XXI. Uluslararası Yapı ve Yaşam Kongresi Bildiriler Kitabı, TMMOB Mimarlar Odası, 20–21 Mart 2009, Bursa.
http://acikerisim.fsm.edu.tr:8080/xmlui/bitstream/handle/11352/2661/Burkut.pdf?sequence=1&isAllowed=y (accessed May 21, 2019).
http://koycegiz.bel.tr (accessed May 20, 2019).
http://mudurnu.bel.tr/ (accessed May 18, 2019).
http://urfafx.mekan360.com/iys_ilcelerimiz,sehirID=63,kriter39=52,icerik=518,sayfa=1-ilcelerimiz-halfeti-ekonomi.html? (accessed May 21, 2019).
http://urfafx.mekan360.com/iys_ilcelerimiz,sehirID=63,kriter39=52,icerik=518,sayfa=1-ilcelerimiz-halfeti-ekonomi.html? (accessed May 14, 2019).
http://wowturkey.com/forum/viewtopic.php?t=3173 (accessed May 02, 2019).
http://www.ahlat.bel.tr/sosyo-ekonomik-durum (accessed May 20, 2019).
http://www.cittaslowgerze.com/gerzeninEkonomisi.aspx (accessed May 21, 2019).
http://www.goynukkaymakamligi.gov.tr/goynuk-ekonomisi (accessed May 11, 2019).
http://www.hendesedergisi.com/yazardetay/27-65-turkiyenin-yavas-sehri-tarakli.aspx (accessed Jun 03, 2019).

http://www.hendesedergisi.com/yazardetay/9-66-cittaslow-dunyanin-sakin-sehir-hareketi.aspx (accessed Jun 10, 2019).
http://www.ispartakulturturizm.gov.tr/TR-165533/egirdir.html (accessed Jun 15, 2019).
http://www.yenipazar.bel.tr/yenipazar-hakkinda (accessed Jun 17, 2019)
https://cittaslowturkiye.org/cittaslow-akyaka/ (accessed May 21, 2019).
https://cittaslowturkiye.org/cittaslow-savsat/ (accessed Apr 21, 2019).
https://cittaslowturkiye.org/cittaslow-savsat/ (accessed Apr 18, 2019).
https://cittaslowturkiye.org/cittaslow-seferihisar/ (accessed May 18, 2019).
https://cittaslowturkiye.org/cittaslow-uzundere/ (accessed May 04, 2019).
https://e-gokceada.com.tr/ada-ekonomisi-ve-uretim/ (accessed May 21, 2019).
https://seferihisaradair.com/seferihisar-hakkinda/ (accessed May 21, 2019).
https://www.ntv.com.tr/ekonomi/turkiye-sakin-sehir-fikrini-sevdi,56-nNdSH5UK0U7ohQ6FkUQ (accessed Jun 13, 2019).
https://www.trakyanet.com/trakya/kirklareli/vize/169-sosyal-ve-ekonomik-durum.html (accessed May 21, 2019).

CHAPTER 6

THE IMPORTANCE OF LOCAL PRODUCTS RELATED TO TOURISM: AN ANALYSIS OF TURKEY'S GEOGRAPHIC INDICATOR MAP OF LOCAL ATTRACTIONS

H. KOSKER* and Z. KURTULAY

Department of Tourism Management, Kdz. Eregli Tourism Faculty, Kdz. Eregli, Zonguldak

**E-mail: hasankosker@yahoo.com; hasankosker@beun.edu.tr*

ABSTRACT

Nowadays, with the development of communication technologies, it is easier for individuals to reach the products they want. This situation increases the competition among enterprises. Therefore, businesses produce different and unique products that can attract consumers. Consumers' preferences change accordingly and they want to be a part of daily life when they go to places they have never been before. In other words he wants to gain experience. Unlike other sectors, tourism is an industry that sells experience to individuals. Businesses that want to take part in the tourism market should also take into account the changing needs of consumers. Therefore, geographic indication, which states that a product is unique to that region, is becoming increasingly important for sectors such as the tourism sector, which offers experience to consumers. This study examined the products with geographical indication or application process in Turkey and Turkey's geographical indication products have been tried to create the map. As a result of study, the number of registered geographical indication products in Turkey are a total of 386 units. 59.84% (231) of the registered products are food, 17.62% (68) are handicrafts, 16.58% (64) are agriculture, 2.33% (9) are industry, 2.07% (8) are mines, 1.29% (5) are animals, and 0.02% (1) are others. There is no city that does not have geographical indication in Turkey. The provinces with the most geographically indicated products are Şanlıurfa (26), Gaziantep (22), İzmir (17), Kahramanmaraş (13), Afyon, Balıkesir, Kayseri (11), and Mersin (10), respectively. The total number of geographical indication products in the application process in Turkey is 411. 67.15% (276) of these products are food, 23.11% (95) were agriculture, 5.59% (23) are handicrafts, 2.67% (9) are mine, 1% 21 (5) are industry and 0.02% (1) are others. The number of products in the geographic indication application process is higher than the products with geographic indication. This may prove to be an increasing awareness of geographical indication.

6.1 INTRODUCTION

Eating and drinking, which is one of the most basic physiological needs in Maslow's Hierarchy of Needs, has been an important factor since the beginning of humanity. Since the prehistoric times, there have been many written documents about ceremonies, feasts, or banquets. In these documents, it is possible to see social, cultural, and religious factors that lead people to eat and drink. Together with the progress of time, these factors changed. People are now eating and drinking because of motivation factors in higher levels of Maslow's Hierarchy of Needs, such as curiosity, experience, self-realization, or an acceptance of others. This situation affects tourism, as well as, every sector where the human element exists.

In today's modern world, consumption patterns of individuals are changing rapidly. Accordingly, the expectations of tourism products are varying. The global tourism market differs from the classical sea–sand–sun activities and with this process individuals' demands for local and natural products are increasing. When tourists visit somewhere new, where they have never been there before, they look for different experiences and wish to be part of the local people's daily lives. For this, they want to experience local products, which are cultural and regional.

Local products play an important role in the tourism market and increase the country's tourism potential (Küçükaltan, 2009, p. 8). Today, countries, which request to take part in the tourism market, are trying to stand out at the forefront with features that distinguish them from other destinations. For this, countries must introduce their local products, which are the most difficult to imitate as a result of hundreds of years of cultural effects. Considering that tourism activities are aimed at obtaining experience, local products are an important part of that experience for tourists. In the modern world, where globalization is increasing, enterprises are able to introduce local products to the global tourism market and carry out the activities necessary for the promotion of the region/country. Globalizing tourism markets make the local products and cultures open to external influences and these local products and cultures can sometimes be threatened by other local products or cultures (Yurtseven et al., 2010).

This situation requires that local products be protected by legal regulations. In the global tourism market competition is escalating; alternative products are increasing and the borders are disappearing day by day. To use local products as a promotion tool and protect these local products, a geographic attraction is needed. Geographic attraction has an important role in continuing the creative efforts and gaining legal ground to protecting the local values for future generations.

Currently, with the development of communication technologies, it is easier for individuals to obtain the products they want. This increases competition among enterprises. Businesses produce different and unique products that can be attractive to consumers. Consumers' preferences change. Taking that into consideration, consumers want to be a part of daily life when they travel to places they have never been to before. In other words, they want to gain new experiences. Unlike other sectors, tourism is an industry that offers change. Businesses, that want to take part in the tourism market, should also take into account the changing needs of consumers. Geographic attractions, unique to that region, are becoming increasingly important to the tourism industry. The purpose of this study is to explain the concept of geographic attractions; to group the local products with a geographic attraction; process the application process and create the map of Turkey's geographical indicated local products. From this viewpoint, as well as, the list of Turkey's geographic indications and statistical data published by the

Turkish Patent Institute are examined and the results are subjected by content analysis.

6.2 GEOGRAPHIC PRODUCT ATTRACTIONS

Global standardization has reduced the variety of products and has led to the emergence of specific products preferred in mass consumption. As a result of this consumption which continued until the demand for different and local products decreases. Thereafter, the demand for natural and local products increased due to crisis such as hormones in hens and transgenetic products (Tekelioğlu, 2010, p. 118). Today's consumers have become more aware and are now more selective about the products they consume, where the products come from, and how those products are produced. This has led people to consume more trustworthy products that are registered by the authorities. Thus, the importance of geographic indicators is enhanced.

The ability to sell a product depends on its recognition by consumers and if it is not recognized, cannot be sold at a level of a high-quality standard. (Shapiro, 1983, p. 497). Geographic awareness is an important element to increase product awareness among manufacturers, sellers, and consumers. Consumers are ready to pay more for products with specific standards and geographic attractions.

Geographic indicating has a long history. Geographic indicating was used to determine the origin of the bricks and stones, which were preferred due to their durability and were used in the construction of pyramids built in Ancient Egypt. In ancient Greece, indicating was used as a sign of the quality of wines from Thasos Island. The first example of geographic indicating appears in France in 1666 in the parliamentary decision to punish counterfeit cheese producers. Today France is a very successful country in geographic indicating. A total of 779 products, including 610 of wine and 169 of agricultural products are under protection (Tekelioğlu, 2010, p. 119).

Geographic indication is a result of the identification of a product within a region or country as a result of nature, reputation, or other characteristics of a product (Nazlı, 2011, p. 3). There are some features that indicating must have to identify as a geographic indication. These are product, region, sign, and the relationship between a particular characteristic of the product and its geographical origin (Kan et al., 2008). Turkish Patent and Trademark Authority (2019, p. 1) defines the concept of geographic indicating as "It refers to the name of a local product which is different from its counterparts and owes this difference to its region. In this sense, the geographic indicating is a sign that a product is identified with the region, area or country and has quality, reputation, and other positive characteristics." The concept of geographic indicating has been the subject of many regulations, but this concept was first used in the European Union Council Regulation in 1992 and Trade Related Intellectual Property Agreement (TRIPS) in 1995 (Barham, 2003; Gündoğdu, 2006).

The concept of geographical indications can be examined in two different groups as *origin* and *merchandise mark*. The difference between the two groups arises from the place of production. Turkish Patent and Trademark Authority (2019, p. 1), defines the concept of origin as "In a region, or an exceptional situation where geographic boundaries are determined, the names of all or all of the main characteristics of this geographical area, which are natural and human". Some products cannot be manufactured somewhere else from their regions. Because the products get their characteristics only in the region in which they are made such as Finike Orange, Malatya Apricot, or Edremit Olive Oil (OKA, 2019, p. 7).

Turkish Patent and Trademark Authority (2019, p. 1) defines the concept of merchandise mark as "names of products, whose production, processing, and other transactions are identified within the boundaries of a defined geographic area, identified with a geographical character, reputation, or other characteristics of a geographical area, region or country." Merchandise products can be manufactured in any other regions but at least one of the product characteristics must be linked to the originated region such as Antep Baklava or Hereke Silk Carpet (OKA, 2019, p. 7). The European Union, by tripartite distinction, classifies geographical indications as origin, merchandise mark, and traditional products (Nazlı, 2011, p. 3). In Turkish law, the basic regulation on geographical indications is the Decree Law No. 555 adopted in 1995. In accordance with Article 3 of the Law, geographic indicating represents a product identified within a region, area or country with a particular quality, reputation, or other characteristics (Bozgeyik, 2009, p. 12).

6.2.1 IMPORTANCE AND AIM OF GEOGRAPHICAL INDICATIONS

Geographical indicators are very important for development and agriculture-based economies. The reason for this is that these countries have not developed an industry and traditional methods in agricultural production are still being maintained (Yalçın, 2013, p. 208). Geographic indicators also do not depend on a single person and therefore do not belong to a single manufacturer. It also protects the rights of persons who produce and/or market products that qualified well enough to receive geographic indicator certificates. Thanks to the geographic indications, local products gain the feature of being attractive to tourists, as well as, being protected (Orhan, 2010).

There are other advantages of indicating local products geographically. These are (Yenipınar et al., 2014, p. 14):

- Protection of the quality of the local products indicated as geographical and to ensure the continuity of this production.
- Ensure that local manufacturers who produce geographically indicated products will benefit from the protection provided by registration.
- Distinguish the counterfeit products that do not have the necessary requirements for geographical indicator certificates, which ensure the protection of the consumer.
- Ensure the protection of the national and cultural values of the region and the country.

By using geographic indicator products, producer, consumer, and region/country will be protected; they can rely on registered products being available to consumers. Although the products accepted as geographic indicators are predominantly food and beverages (Gökovalı, 2007, p.142); all cultural values, for example, in handicrafts, traditional events (birth, marriage, and death), religious places, and ceremonies can be considered as local products (Lanquar, 1995, p. 53).

Geographic indicating aims to document the usage rights of local products. Other important purposes are to ensure the production of products and to maintain quality, to protect the rights of producers in the region and have recognition at national and international levels. Thus, the recognition of the traditional and local products by the next generations will be guaranteed, and the demand for those products will be increased (Karakulak, 2016, p. 7).

Rights that geographic indicators provide to manufacturers, consumers, regions, and countries are as follows (Turkish Patent and Trademark Authority, 2019, p. 3):

- Manufacturers who have the right to use geographical indications may demand sanctions against third parties if they violate their rights.
- Persons wishing to benefit from the reputation of the geographical indications may be prevented by the court from doing so.
- Misleading use of similar products can be prevented by the court.
- The misleading use of product emblems/symbols with a geographic indicator can be prevented by the court.

Geographic indicators have a good protection system and can be a very efficient marketing tool with great economic value if they are used for appropriate purposes (Addor and Grazioli, 2002). If these local products are not protected, they can be exposed to commercial abuse. Consumers' confidence for these products decreases. For example, the amount of coffee produced in the Antigua region of Guetamala is three thousand tons. However, 25 thousand tons of coffee is sold in the world under this name. Similarly, while 10 thousand tons of Darjeeling tea is produced in India, the amount of tea sold in the world is around 30 thousand tons (Tekelioğlu, 2010, p. 119).

6.3 GEOGRAPHICAL INDICATIONS AND TOURISM

It is understood that a tourism sector is an important tool for marketing natural, cultural, and historical resources, bridging the cultures and promoting cultural and traditional values of countries. One of the most important values is local products. Today many products such as food, clothes, or handicrafts are manufactured in a global market and many of these products are known with the same? Name of the area where it is produced. The products, whose fame and quality are taken from natural or human resources of their location are shaped by knowledge, skill, experience, and tradition. They are local products (Tekelioğlu and Demirer, 2008, p. 87; Vittori, 2010, p. 304).

Products that are registered with a geographic indicator can make the area, where they are produced, and a center of attraction. A geographic indicator is a guarantee of the standards and quality of products. In some cases, the local products are the result of tourist influences. Local products are important for national and international tourism (McKrecher et al., 2008, p. 138; Yüncü, 2009, p. 29; Altınel, 2009, p. 15). All-inclusive service and sea–sand–sun-based tourism which many hotels provide today has become customary for tourists. At the present time tourists have become more curious; request to experience the differences and demand to make contact with local people. Therefore, tourists prefer rural and natural places and local products.

Every country's cuisine is nourished by its own culture and shaped according to its habits and desires. The preparation, cooking, serving and storage of food and beverages, and the tools and equipment required for this are varied (Kesici, 2012, p. 34). Regions and countries have different cultural elements from their competitors. Cuisines of the regions and countries have important potential for marketing tourism (Winter et al., 2008; Richards, 2012; Gökovalı, 2007). For example, Italy promotes not only its architectural and artistic heritage but also its cuisine. Today, Italian cuisine is remembered with pizza and pasta varieties. France is identified with wine and has become worldwide famous for its Bordeaux and Bourgogne wines. In Spanish cuisine, paella and tapas come to the forefront and sushi has an important place in Far East cuisine (Gurme Guide, 2019).

Due to its geographical location, the culture and historical background of Turkey, there is a specific culinary culture and a large variety of products. During the historical period, the Turks who live in Central Asia had a simple food culture. After the resident life, Turkish cuisine in the Seljuk and Ottoman period had glamorous flavors. Today, Turkish cuisine has a combination of different tastes and has become very popular all around the World (Durlu-Özkaya and Can, 2012, p. 28). Considering that one-third of the tourist expenses are made for food and beverages (Mak et al., 2012, p. 172; Tikkannen, 2007, p. 721), the importance of local products for tourism might be understood.

Due to the understanding of the importance of geographic indicators for the tourism industry, it has recently become a frequently researched topic. When the studies are examined, it is observed that geographic indicators, generally examine and appraise the socio-economic and cultural dimensions, the protection of local products and sustainability, regulation in national and international law, and the impact on consumer preferences and tourism (Addor and Grazioli, 2002; Albayrak and Güneş, 2010; Arfini et al., 2011; Bowen and Zapata, 2009; Bozgeyik, 2009; Caenegem et al., 2015; Demirer, 2010; Doğan and Gökovalı, 2012; Ilbert and Petit, 2009; Kan et al., 2008, 2010; Marzella, 2008; Polat, 2015; Suh and MacPherson, 2007; Treager, 2003).

6.4 METHODOLOGY

The purpose of this study is to categorize local products that have a geographic indicator or an application process for it, to detailing products as main and subgroups, and to create a map of Turkey's local products. For this purpose, the list of Turkey's geographic indicators and statistical data published by the Turkish Patent Institute are examined by content analysis. The main purpose of content analysis is to reach the concepts and relations that can explain the collected data. Data is analyzed through content analysis and then revealed (Yıldırım and Şimşek, 2011, p. 39).

Before local products had a geographic indicator analyzed, the products originating from Italy, France, Scotland, and the Turkish Republic of Northern Cyprus were eliminated from the product list. Then the products were classified and grouped. Before the basic and detailed product group is registered with a geographic indicator, their grouping is made according to the specified basic product and detailed product group. However, many of the products in the geographic indicator application process have not been detailed. Agriculture and food groups are intertwined and it has been noticed that a product that should be in a handicraft group is mistakenly in the food group.

In order to solve this confusion and to make a reliable analysis, the products in the application process are divided into a main and subgroup groups. This process is done taking into consideration, the grouping of registered products. Five products in two separate institutions in the same city were counted twice and these products were account once. After, the products with geographic indicators are divided into seven main groups: food, agriculture, handicrafts, industry, animal, mining and others.

The food group was divided into 13 subgroups, agriculture was divided into two subgroups, and handicraft was divided into five subgroups. In Table 6.1, frequency summaries of main and subgroups product which have a geographic indicator or are in the application process are calculated. In Table 6.2, frequency totals of main and subgroup products, which have a geographic indicator or are in the application process in Turkey, are calculated by general and provincial basis.

6.5 FINDINGS

The findings obtained from the analysis are given in the Table 6.1.

Turkey's cultural history is effective in countries with agricultural and livestock groupings of geographical products. Table 6.1 shows the distribution of geographic registered indicators and are in the application process by product groups. These products are classified into seven main groups: food, agriculture, handicrafts, industry, animal, mining, and others. Food itself is divided into 13 groups, agriculture 2, handicrafts 5. The number of registered geographic indication products in Turkey is a total of 386 units. Food ranks first as the most registered product group. This is followed by handicrafts and agricultural products. 59.84% (231) of the registered products are food, 17.62% (68) are handicrafts, 16.58% (64) are agriculture, 2.33% (9) are industry, 2.07% (8) are mines, 1.29% (5) are animals, and 0.02% (1) are others.

The most registered products in the food group are meals and soups, bakery products, pastries and desserts, chocolate candies and derivatives, processed and unprocessed fruit, and vegetables and mushrooms. The least registered ones are cheese and nonbutter dairy products, condiments for meals, flavors, sauces and salt, ice cream and edible salts, and soft drinks. The most registered products in the handcraft group are carpets and rugs and weavings. The least registered are gold, copper and other types of metal working, walking stick and other types of woodwork. The most registered agricultural products are processed and unprocessed fruits and vegetables and mushrooms.

The total number of geographic products in the application process in Turkey is 411. 67.15% (276) of these products are food, 23.11% (95) are agriculture, 5.59% (23) are handicrafts, 2.67% (9) are mine, 1.21% (5) are industry, and 0.02% (1) are others. Here, food is the most frequently

TABLE 6.1 Distribution of Geographical Indications in Registered and Application Process by Product Groups.

Main production group	Registered geographical indication product		Products in application process of geographical indication	
	Total	%	Total	%
Food	**231**	**59.84**	**276**	**67.15**
Meals and soups	92	39.82	61	22.10
Cheeses	13	5.63	21	7.61
Butter, fats, and oils	9	3.90	4	1.45
Cheese and nonbutter dairy products	2	0.87	9	3.26
Bakery products, pastries, and dessert	50	21.6	54	19.56
Processed and unprocessed fruit, vegetables, and mushrooms	20	8.66	45	16.30
Condiments for meals, flavors, sauces, and salt	2	0.87	7	2.57
Processed and unprocessed meat products	6	2.59	9	3.26
Chocolate candies and derivatives	22	9.52	34	12.32
Ice cream and edible salts	2	0.87	2	0.72
Honey	4	1.73	14	5.07
Soft drinks	2	0.87	15	5.43
Beer and other alcoholic drinks	6	2.60	1	0.36
Agriculture	64	16.58	95	23.11
Processed and unprocessed fruit, vegetables, and mushrooms	61	95.31	79	83.16
Others	3	4.69	16	16.84
Handicraft	68	17.62	23	5.59
Carpets and rugs	28	41.18	7	30.44
Weavings	20	29.41	9	39.13
Walking stick and other types of woodworking	3	4.41	1	4.35
Gold, copper and other types of metal working	4	5.88	–	–
Others	13	19.11	6	26.08
Industry	9	2.33	5	1.21
Animals	5	1.29	–	–
Mines	8	2.07	11	2.67
Others	1	0.02	1	0.02
General total	386	100.0	411	100.0

Source: Adapted from Turkish Patent and Trademark Authority (2019).

applied product group. This is followed by agricultural products and handicrafts. The products which are most applied for are the food group; food and soups, bakery products, pastries and desserts, processed and unprocessed fruits and vegetables and mushrooms, as well as chocolate candies and derivatives.

Those who are at least in the application stage are beers and other alcoholic beverages; fats and oils including ice cream and edible salts, and butter. The products that are in the most application process within the agricultural group are processed and unprocessed fruits and vegetables and mushrooms. The products which are at the most application stage in the handicraft group are carpets and rugs. The ones that are at least in the application stage are walking stick and other wood processing types.

In order to see the product groups within the scope of geographic indication and the application process, percentage (%) distribution of product groups is shown on the chart. Figure 6.1 shows the percentage (%) of registered product groups and Figure 6.2 shows the percentage (%) of product groups in the application process. In both figures, it is noteworthy that the food group is far ahead. Figure 6.1 shows food products, handicrafts, and agricultural products. In Figure 6.2, food products followed by agricultural products and handicrafts.

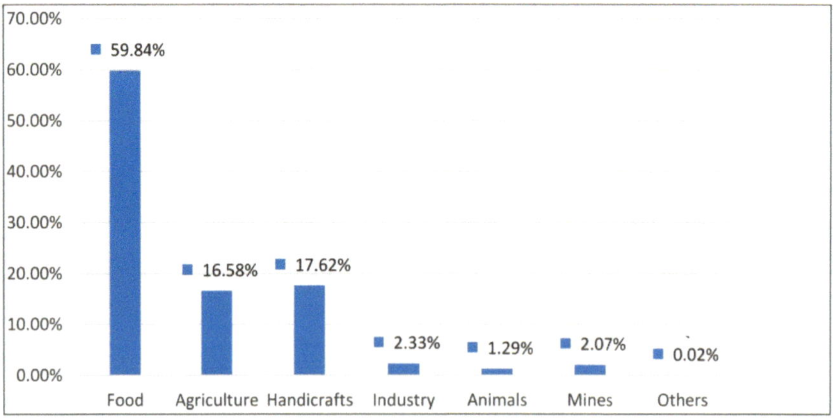

FIGURE 6.1 Proportion of registered geographical indication products by groups.
Source: Adapted from Turkish Patent and Trademark Authority (2019).

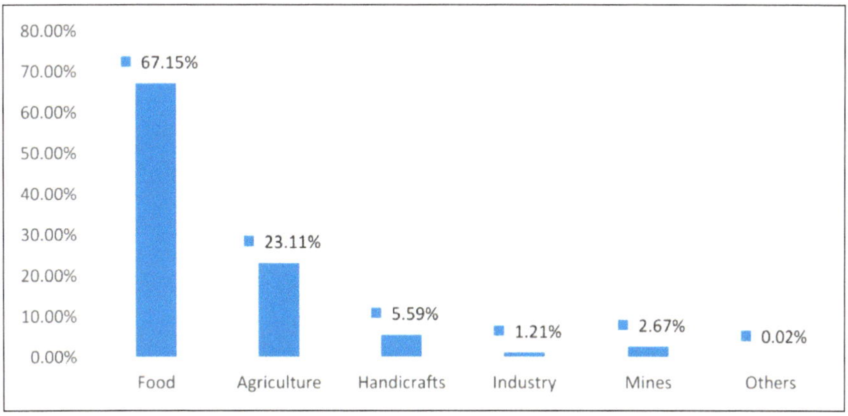

FIGURE 6.2 Proportion of products in geographical indication application process according to groups.
Source: Adapted from Turkish Patent and Trademark Authority (2019).

Table 6.2 shows the detailed distribution of geographically marked products by provinces and product groups of Turkey's proprietary products and the ones in the application process. There are no provinces with any geographically marked products. Each province has at least one geographically marked product. The provinces with the most geographically marked products are Şanlıurfa (26), Gaziantep (22), İzmir (17), Kahramanmaraş (13), Afyon, Balıkesir and Kayseri (11), and Mersin (10), respectively. The provinces with the least registered geographically marked product are Adana, Ağrı, Artvin, Bilecik, Bingol, Bitlis, Çankırı, Istanbul, Kırklareli, Mus, Tekirdağ, Tunceli, Zonguldak, Aksaray, Karaman, Batman, Şırnak, Bartın, Yalova, and Osmaniye. The number of products in the geographic indication application process is higher than the products with geographic indication. This may prove to be an increasing awareness of geographic indication. The provinces that applied for the most geographical indications are Sivas (29), Adana (21), Bolu and Gaziantep (20), Kastamonu (16), Muğla (12), Konya (11), and Mersin (10), respectively. It is noteworthy that Istanbul, Turkey's cultural capital attracts most tourists and has only one product with a geographical indication and no product in the application process. It is understood that the public and nongovernmental organizations in Istanbul neglect the subject of geographical attraction.

TABLE 6.2 Distribution of Geographically Marked Products by Provinces and Product Groups of Turkey's Proprietary Products and the Ones in Application Process.

Number	Province name	Number of geographical indication product								Number of products in application process of geographical indication							
		Food	Agriculture	Handicrafts	Industry	Animals	Mines	Others	Total	Food	Agriculture	Handicrafts	Industry	Animals	Mines	Others	Total
1	Adana	1	–	–	–	–		–	**1**	14	6	1	–		–	–	21
2	Adıyaman	2	1	–	–	–	–	–	**3**	1	2	1	–	–	–	–	4
3	Afyon	5	1	4	–	–	1	–	11	2	3	1	–	–	–	–	6
4	Ağrı	1	–	–	–	–	–	–	**1**	1	–	–	–	–	–	–	1
5	Amasya	1	2	–	–	–	1	–	**4**	–	1	–	–	–	–	–	1
6	Ankara	5	1	2	–	–	1	–	**9**	7	1	–	1	–	–	–	9
7	Antalya	1	5	1	–	–	–	–	7	5	2	–	–	–	–	–	7
8	Artvin	1	–	–	–	–	–	–	**1**	3	–	–	–	–	–	–	3
9	Aydın	2	2	1	–	–	–	–	**5**	7	–	1	–	–	–	–	8
10	Balıkesir	6	2	2	–	–	1	–	**11**	1	–	–	–	–	–	–	1
11	Bilecik	1	–	–	–	–	–	–	1	3	3	–	–	–	–	–	6
12	Bingöl	1	–	–	–	–	–	–	**1**	2	–	–	–	–	–	–	2
13	Bitlis	–	–	1	–	–	–	–	**1**	2	–	1	–	–	–	–	3
14	Bolu	1	1	1	–	–	–	–	**3**	13	6	–	–	–	1	–	20
15	Burdu	2	1	–	–	–	–	–	3	3	–	–	–	–	–	–	3

TABLE 6.2 *(Continued)*

Number	Province name	Number of geographical indication product								Number of products in application process of geographical indication							
		Food	**Agriculture**	**Handicrafts**	**Industry**	**Animals**	**Mines**	**Others**	**Total**	**Food**	**Agriculture**	**Handicrafts**	**Industry**	**Animals**	**Mines**	**Others**	**Total**
16	Bursa	3	2	1	1	1	–	–	**8**	4	1	–	1	–	–	–	6
17	Çanakkale	1	2	1	–	–	–	–	**4**	6	–	–	–	–	–	–	6
18	Çankırı	1	–	–	–	–	–	–	**1**	4	–	–	–	–	–	–	4
19	Çorum	2	–	–	–	–	–	–	2	2	–	–	–	–	–	–	2
20	Denizli	2	3	1	–	–	1	–	**7**	4	–	–	–	–	–	–	4
21	Diyarbakır	3	–	–	–	–	–	–	**3**	2	1	–	2	–	1	–	6
22	Edirne	3	1	–	–	–	–	–	**4**	1	–	1	–	–	–	–	2
23	Elazığ	2	1	–	–	–	–	–	3	3	1	–	–	–	1	–	5
24	Erzincan	2	–	1	–	–	–	–	**3**	1	2	–	–	–	–	–	3
25	Erzurum	5	2	1	–	–	1	–	**9**	6	1	–	–	–	–	–	7
26	Eskişehir	3	–	–	–	–	1	–	**4**	1	–	1	–	–	–	–	2
27	Gaziantep	14	3	4	1	–	–	–	22	13	7	–	–	–	–	–	20
28	Giresun	2	2	–	–	–	–	–	**4**	5	2	–	–	–	–	–	7
29	Gümüşhane	4	–	–	–	–	–	–	**4**	4	1	1	–	–	–	–	6
30	Hakkari	3	–	–	–	–	–	–	**3**	–	–	–	–	–	–	–	0
31	Hatay	3	–	–	2	–	–	–	5	3	2	–	–	–	–	–	5

TABLE 6.2 *(Continued)*

Number	Province name	Number of geographical indication product								Number of products in application process of geographical indication							
		Food	Agriculture	Handicrafts	Industry	Animals	Mines	Others	Total	Food	Agriculture	Handicrafts	Industry	Animals	Mines	Others	Total
32	Isparta	3	1	2		–	–	–	**6**	3	1	–	1	–	–	–	5
33	Mersin	5	5	–	–	–	–	–	**10**	5	5	–	–	–	–	–	10
34	İstanbul	–	–	1	–	–	–	–	**1**	–	–	–	–	–	–	–	0
35	İzmir	8	6	2	–	–	–	1	**17**	4	4	–	–	–	–	–	8
36	Kars	2	1	2	–	1	–	–	**6**	–	–	–	–	–	–	–	0
37	Kastamonu	5	1	1	–	–	–	–	**7**	15	–	1	–	–	–	–	16
38	Kayseri	6	2	3	–	–	–	–	**11**	5	–	–	–	–	–	–	5
39	Kırklareli	1	–	–	–	–	–	–	**1**	3	–	–	–	–	–	–	3
40	Kırşehir	2	–	–	–	–	–	–	**2**	–	–	–	–	–	–	–	0
41	Kocaeli	1	–	5	–	–	–	–	**6**	7	–	–	–	–	–	–	7
42	Konya	3	2	–	–	–	–	–	**5**	7	4	–	–	–	–	–	11
43	Kütahya	1	–	2	–	–	–	–	**3**	1	1	–	–	–	–	–	2
44	Malatya	–	4	–	–	–	–	–	**4**	5	2	1	–	–	–	–	8
45	Manisa	5	1	3	–	–	–	–	**9**	5	3	–	–	–	–	–	8
46	K. Maraş	6	2	4	1	–	–	–	**13**	4	3	–	–	–	–	–	7
47	Mardin	5	–	–	–	–	1	–	**6**	1	–	–	–	–	–	–	1

TABLE 6.2 *(Continued)*

Number	Province name	Number of geographical indication product								Number of products in application process of geographical indication							
		Food	Agriculture	Handicrafts	Industry	Animals	Mines	Others	Total	Food	Agriculture	Handicrafts	Industry	Animals	Mines	Others	Total
48	Muğla	6	–	1	–	–	–	–	**7**	2	6	2	–	–	2	–	12
49	Muş	1	–	–	–	–	–	–	**1**	2	2	–	–	–	–	–	4
50	Nevşehir	1	–	1	–	–	–	–	**2**	–	1	–	–	–	–	–	1
51	Niğde	1	–	1	–	–	–	–	**2**	–	1	–	–	–	1	–	2
52	Ordu	3	–	–	–	–	–	–	**3**	2	–	–	–	–	2	–	4
53	Rize	2	–	1	–	–	–	–	**3**	6	1	–	–	–	–	–	7
54	Sakarya	2	–	–	1	–	–	–	**3**	1	2	–	–	–	–	–	3
55	Samsun	7	–	1	1	–	–	–	**9**	4	–	2	–	–	–	–	6
56	Siirt	3	–	1	–	–	–	–	**4**	–	–	1	–	–	–	–	1
57	Sinop	1	–	3	–	–	–	–	**4**	3	1	2	–	–	–	–	6
58	Sivas	2	1	1	1	2			7	22	3	4	–	–	–	–	29
59	Tekirdağ	1	–	–	–	–	–	–	**1**	3	–	1	–	–	–	–	4
60	Tokat	4	2	–	–	–	–	–	**6**	2	–	1	–	–	–	–	3
61	Trabzon	4	–	3	1	–	–	–	**8**	2	–	–	–	–	–	–	2
62	Tunceli	1	–	–	–	–	–	–	**1**	–	2	–	–	–	–	–	2
63	Şanlıurfa	25	1	–	–	–	–	–	**26**	4	2	–	–	–	–	–	6

TABLE 6.2 *(Continued)*

Number	Province name	Number of geographical indication product								Number of products in application process of geographical indication							
		Food	Agriculture	Handicrafts	Industry	Animals	Mines	Others	Total	Food	Agriculture	Handicrafts	Industry	Animals	Mines	Others	Total
64	Uşak	1	–	2	–	–	–	–	**3**	–	–	–	–	–	–	–	0
65	Van	4	–	2	–	–	–	–	**6**	3	–	–	–	–	–	–	3
66	Yozgat	4	–	–	–	–	–	–	**4**	–	–	–	–	–	–	–	0
67	Zonguldak	–	–	1	–	–	–	–	**1**	4	–	–	–	–	–	–	4
68	Aksaray	–	–	1	–	–	–	–	**1**	3	3	–	–	–	–	–	6
69	Bayburt	2	–	–	–	–	–	–	**2**	3	–	–	–	–	1	–	4
70	Karaman	1	–	–	–	–	–	–	**1**	1	–	–	–	–	–	–	1
71	Kırıkkale	1	1	–	–	–	–	–	**2**	–	2	–	–	–	–	–	2
72	Batman	1	–	–	–	–	–	–	**1**	–	1	–	–	–	–	–	1
73	Şırnak	1	–	–	–	–	–	–	**1**	–	–	–	–	–	–	–	1
74	Bartın	–	–	1	–	–	–	–	**1**	6	–	–	–	–	–	–	6
75	Ardahan	1	–	1	–	–		–	**2**	1	–	–	–		–	–	1
76	Iğdır	2	–	–	–	–	–	–	**2**	1	–	–	–	–	–	–	1
77	Yalova	–	–	1	–	–	–	–	**1**	1	–	–	–	–	–	1	2
78	Karabük	2	1	1	–	–	–	–	**4**	5	1	–	–	–	2	–	8
79	Kilis	2	–	–	–	–	–	–	**2**	–	–	–	–	–	–	–	0

TABLE 6.2 *(Continued)*

Number	Province name	Number of geographical indication product								Number of products in application process of geographical indication							
		Food	**Agriculture**	**Handicrafts**	**Industry**	**Animals**	**Mines**	**Others**	**Total**	**Food**	**Agriculture**	**Handicrafts**	**Industry**	**Animals**	**Mines**	**Others**	**Total**
80	Osmaniye	–	1	–	–	–	–	–	**1**	1	–	–	–	–	–	–	1
81	Düzce	4	–	–	–	–	–	–	**4**	4	2	–	–	–	–	–	6
	Türkish	6	–	–	–	1	–	–	7	2	–	–	–	–	–	–	2
General Total		**231**	**64**	**68**	**9**	**5**	**8**	**1**	**386**	**276**	**95**	**23**	**5**	**0**	**11**	**1**	**411**

Source: Adapted from Turkish Patent and Trademark Authority (2019).

6.6 CONCLUSION

In the growing tourism market, geographic attractions are used as a commercial tool and become an important source of economic income for the local people. In developing countries, with economies based on agriculture, most of their exports are local products such as food, clothing, or handicrafts. If a product is a geographic indicator, it has a standardized quality and is created within the local culture.

Increasing population, changing socio-economic structure, and technological opportunities have caused changes in human needs and consumption habits. As a result of globalization, standardized products cannot fulfill the needs of consumers. The crisis in the food sector has encouraged people to consume natural and local products.

Tourism is one of the sectors affected by these changes in the world. Experiencing the differences and the culture of the local people and wanting to be a part of it have changed the tourism demands. Today, tourists prefer products with the characteristics of the region where they visit. This situation increases the importance of geographic products. Geographic products are a guarantee of standardized quality and natural and regional production. Turkey's location and cultural heritage afford great potential for the tourist industry.

The number of registered geographic product attractions in Turkey is 386 units. Food ranks first as the most registered product group. This is followed by handicrafts and agricultural products. The most registered products in the food group are meals and soups, the least registered products are cheese and nonbutter dairy products, condiments for food, flavors, sauces, and salt, ice cream and edible salts, and soft drinks. The most registered products in the handicraft group are carpets and rugs. The least registered are walking sticks and other woodworks. The most registered agricultural products are processed and unprocessed fruits and vegetables and mushrooms. The total number of geographical indication products in the application stage in Turkey is 411 pieces. Here, food is the most attractive indicator. This is followed by agricultural products and handicrafts. The products with the most application stage in the food group are meals and soups. Those that are the least in the application stage are beers and other alcoholic beverages; fats and oils including ice cream and edible salts and butter. The products that are in the most application process within the agricultural group are processed and unprocessed fruits and vegetables and mushrooms. The products which are at the most application stage in the handicraft group are carpets and rugs. The ones that are at least in the application stage are walking stick and other wood processing products.

There are no provinces without geographically marked products in Turkey. Each province has at least one geographically marked product. The provinces with the most geographically marked products are Şanlıurfa (26), Gaziantep (22), İzmir (17), Kahramanmaraş (13), Afyon, Balıkesir and Kayseri (11), Mersin (10), respectively. The provinces with the least geographically marked product are Adana, Ağrı, Artvin, Bilecik, Bingöl, Bitlis, Çankırı, İstanbul, Kırklareli, Muş, Tekirdağ, Tunceli, Zonguldak, Aksaray, Karaman, Batman, Şırnak, Bartın, Yalova, and Osmaniye. The number of products in the geographical indication application process is remarkable. This may prove to be an increasing awareness of geographical indication. The provinces that applied for the most geographical indication are Sivas (29), Adana (21), Bolu and Gaziantep (20), Kastamonu (16), Muğla (12), Konya (11), and Mersin (10).

KEYWORDS

- **geographical indication**
- **local production**
- **consume**
- **tourism**
- **Turkey**

REFERENCES

Addor, F.; Grazioli, A. Geographical Indications beyond Wines and Spirits: A Roadmap for a Better Protection for Geographical Indications in the WTO TRIPS Agreement. *J. World Intellectual Property* **2002,** *5* (6), 865–897.

Albayrak, M.; Güneş, E. Traditional Foods: Interaction Between Local and Global Foods in Turkey. *Afr. J. Bus. Manage.* **2010,** *4* (4), 555–561.

Altınel, H. *Menu Management in Gastronomy;* Istanbul University Institute of Social Sciences: İstanbul, 2009.

Arfini, F.; Albisu, L. M.; Giacomini, C. Current Situation and Potential Development of Geographical Indications in Europe. In *Labels of Origin for Food. Local Development, Global Recognition*; Barham, E., Sylvander, B., Ed.;. CABI International: Cambridge, USA, 2011.
Barham, E. Translating Terroir: The Global Challenge of French AOC. *J. Rural Stud.* **2003,** *19*, 127–138.
Bowen, S.; Zapata, A. V. Geographical Indications, Terroir, and Socioeconomic and Ecological Sustainability: The Case of Tequila. *J. Rural Stud.* **2009,** *25* (1), 108–119.
Bozgeyik, H. Use and Control of Geographical Indications (On Registration of Antep Baklava as a Geographical Indicator). *FMR* **2009,** *9*, 9–21.
Caenegem, W.; Drahos, P.; Cleary, J. *Provence of Australian Food Products: Is There A Place for Geographical Indications?* RIRDC Publication: Australia, 2015.
Demirer, H. R. *Local Product and Geographical Indications: A Study on France and Turkey;* Akdeniz University Institute of Social Sciences: Antalya, 2010.
Doğan, B.; Gökovalı, Ü. Geographical Indications: The Aspect of Rural Development and Marketing Through the Traditional Products. *Procedia Social Behav. Sci.* **2012,** *62*, 761–765.
Durlu-Özkaya, F.; Can, A. The Effect of Gastronomy Tourism on Destination Marketing. *Turk. Agric. J.* **2012,** *206*, 28–33.
Gökovalı, U. Geographical Indications and Economic Impact: The Case of Turkey. *Muğla Univ. J. Econ. Administrative Sci.* **2007,** *21* (2), 141–160.
Gündoğdu, G. *Concept of Geographical Signage and Protection in Turkish Law*; Beta Printing and Publishing: İstanbul, 2006.
Gurme Guide. *Gurme Guide*. 2019. Retrieved from www.gurmeguide.com.
Ilbert, H.; Petit, M. Are Geographical Indications a Valid Property Right? Global Trends and Challenges. *Dev. Policy Rev.* **2009,** *27* (5), 503–528.
Kan, M.; Gülçubuk, B.; Küçükçongar, M. Possibilities of Use of Geographical Indications in Rural Tourism. *Uludağ Univ. J. Faculty Agric.* **2008,** 57–66.
Kan, M.; Gülçubuk, B., Kan, A.; Küçükçongar, M. Karaman Divle Tulum Cheese As Geographical Indication. *KMU J. Social Econ. Res.* **2010,** *12* (19), 15–23.
Karakulak, Ç. *Gastronomic Identity Creation by Geographical Indicating and The Role of Gastronomic Identity in Destination Marketing: Example of Thrace;* Muğla Sıtkı Koçman University Institue of Social Science: Muğla, 2016.
Kesici, M. The Role of Food and Beverage Culture in Local Demand for Rural Tourism. *KMU J. Social Econ. Res.* **2012,** 33–37.
Küçükaltan, G. *The Effects of Local Tastes on Tourists' Destination Preferences and Country Economies in Globalization Process*. 3rd National Gastronomy Symposium Proceedings, Antalya, 2009.
Lanquar, R. *Tourisme et Environnement en Méditerranéen*; Economica: Paris, 1995.
Mak, A. H.; Lumbers, M.; Eves, A. Globalisation and Food Consumption in Tourism. *Ann. Tourism Res.* **2012,** *39* (1), 171–196.
Marzella, D. A. Culinary Tourism: Does Your Destination Have Potential. *Travel Market. Decis.* **2008,** *9* (2), 2–4.
McKrecher, B.; Okumuş, F.; Okumuş, B. FoodTourism as a Viable Market Segment: It's All How You Cook the Numbers. *J. Travel Tourism Market.* **2008,** *25* (2), 137–148.
Nazlı, C. *Geographical Indications in UN.* 2011. Retrieved from http://abdgm.tarim.gov.tr/yayn_files/yaynlar/11/10.pdf.

OKA. *Middle Black Sea Region Geographically Indicated Products.* 2019. Retrieved from www.oka.org.tr: http://www.oka.org.tr/Documents/OKA_TR83_CI_Katalog_Web_V2.pdf.

Orhan, A. Use of Geographical Indications in Converting Local Values to Tourism Products: The Example of Izmit Refrain. *Anatolia* **2010,** *21* (2), 243–254.

Polat, E. Tourism and Geographical Indication: The Case of Balıkesir. *Meric Int. J. Social Cul. Res.* **2015,** *1* (1), 12–24.

Richards, G. An Overview of Food and Tourism Trends and Policies in OECD. In *Food and the Tourism Experience*; OECD Publishing, 2012.

Shapiro, C. Optimal Pricing of Experience Goods. *Bell J. Econ.* **1983,** *14* (2), 497–507.

Suh, J.; MacPherson, A. The Impact of Geographical Indication on the Revitalization of A Regional Economy: A Case Study of Boseong Green Tea. *Area* **2007,** *39* (4), 518–527.

Tekelioğlu, Y. Regional Products, Geographical Indications and Turkey Practise. *Geographical Indications in Turkey and other Mediterranean countries, Socio-economic Mobility and Bio-cultural Equity*; Antalya, 2010; pp. 118–121.

Tekelioğlu, Y.; Demirer, R. In Globalization Process, Future of Local Products and Geographical Indications. *İGEME's Rev. Mag.* **2008,** (36), 87–102.

Tikkannen, I. Maslow's Hierarchy and Food Tourism in Filand: Five Cases. *Br. Food J.* **2007,** *109* (9), 721–734.

Treager, A. From Stilton to Vimto: Using Food History to Re-think Typical Products in Rural Development. *Sociologia Ruralis* **2003,** *43* (2), 91–107.

Turkish Patent and Trademark Authority. *Geographical Indications and Traditional Product Names Reference Guide.* 2009. Retrieved from https://www.turkpatent.gov.tr/TURKPATENT/commonContent/CAbout.

Vittori, M. The International Debate on Geographical Indications (GIs): The Point of View of the Global Coalition of GI Producers—Origin. *J. World Intellectual Property* **2010,** *13* (2), 304–314.

Winter, T.; Teo, P.; Chang, T. C. *Asia on Tour: Exploring the Rise of Asian Tourism;* Routledge: New York, 2008.

Yalçın, B. Evaluations on the Marketing of Local Products. *Mediterr. Art J.* **2013,** *6* (11), 205–213.

Yenipınar, U., Köşker, H.; Karacaoğlu, S. Importance of Local Foods in Tourism and Geographical Marking: Van Herb Cheese. *J. Tourism Gastronomy Stud.* **2014,** *2* (2), 13–23.

Yıldırım, A.; Şimşek, H. *Qualitative Research Methods in the Social Sciences.* Detay Publishing: Ankara, 2010.

Yurtseven, H. R.; Kaya, O.; Harman, S. *Slow Movement.* Detay Publishing: Ankara, 2010.

Yüncü, H. R. *Gastronomy Tourism and Perşembe Plateau in Terms of Sustainable Tourism.* Aybastı-Kabataş Congress, Eskişehir, 2009.

CHAPTER 7

TOURISTIC PREFERENCES OF CITIZENS RESIDENT ABROAD AND FOREIGNERS: DISCRIMINANT ANALYSIS APPLICATION

N. VATANSEVER TOYLAN[1*] and İ. AKHİSAR[2]

[1]*Faculty of Tourism, Kırklareli University, Turkey*

[2]*Faculty of Business Administration, Kocaeli University, Turkey*

**Corresponding author. E-mail: nilufer.vatansever@klu.edu.tr*

ABSTRACT

The aim of this study is to determine the differences in motivated preferences (choices) for visiting Turkey, between Turkish citizens residing abroad and foreign visitors. Thus, the study will demonstrate the differences between the two groups by determining the purposes of traveling to Turkey. The scope of the research is limited to TurkStat tourist data (Departing Visitors Survey) from tourists that visited Turkey between the years 2009–2018. It is determined by discriminating analyses of how tourists—international tourists visiting Turkey and Turkish people residing abroad are reduced to Turkey's visiting reasons. Thus, it will have been revealed that tourists prefer our country as a touristic destination by which purpose of visit. Linear discriminant analysis is a data discrimination technique that seeks transformation to maximize the ratio of the between class scatter and within class scatter. The method gives the best group of variables for different or same group sizes. In this chapter a variable selection with respect to stepwise procedure in discriminant analysis and applied to the tourism data for the period 2009-2018. The study proves that the model can predict

This study is a renewed version of presented as a paper in 3rd International Congress of Tourism and Management Researches in 2016, in Antalya.

type of touristic preferences for Turkish citizens and foreigners with more than 96% accuracy. These results have important policy implications not only for Turkey, but also for countries and regions around the world has similar socio-economic structure with Turkey.

7.1 INTRODUCTION

Centuries ago, tourism destinations developed around unique natural features such as mineral healing waters, beach and mountain resorts, and sacred sites (Sheldon and Park, 2009). However, international tourism, with its emphasis on the exploitation of "free" resources (sun, see, sand, and friendly people), became an attractive option for economic development for less developed countries (Berno and Bricker, 2001, p 2). Sea, Sun, Sand, and hospitality as attractions for a destination are located in the first place among the reasons of visiting a destination. It is an important information source especially for least developed countries and developing countries, which look at the tourism as an alternative economic source.

The importance of outbound market to a destination country has drawn scholars' interests in understanding tourists' motivation to travel overseas. In this study, various variables are used to discriminate between the purposes of the visit of Turkish citizens who reside abroad and visit Turkey and foreign tourists. Nine variables are determined as leisure, entertainment, sports and cultural activities; visit relatives and friends; education and training; health and medical reasons; religious and pilgrimage; shopping; transit passengers; business and incentive travel and other purposes. Thus, in the study we investigate if these two groups can be separated by the nine variables. At the same time, knowledge of travelers' preferences (purpose of visit) is critical to predict future travel patterns to Turkey.

Discriminant analysis is a technique used to make predictions about the classification of variables. Linear discriminant analysis (LDA) is a well-known classical statistical technique to find the projection that maximizes the ratio of scatter among the data of different classes to scatter within the data of the same class. The univariate approach was later improved and extended to multivariate analysis by Altman (1968) who considered several variables simultaneously using multiple discriminant analysis (MDA). During the years that followed, many researchers attempted to increase the success of MDA in predicting business failure (Dimitras et al.,

1996). Although discriminant analysis is used in a wide range of tourism studies, there is not any research about discrimination of purpose of visit between citizen resident abroad and foreign tourists. Thus, discriminant analysis is used in this study to determine how the two groups are classified by three different travel purposes.

This chapter is structured into three sections. In the first section, tourist motivations and preferences regarding the purpose of travelling abroad are examined through a literature review. In the second section, the methodology of the study is presented. In the third and last section, the results are presented and discussed and their implications for further research are suggested.

7.2 LITERATURE REVIEW

People have tended to travel for different purposes throughout history (Öztürk and Bayat, 2011; Kozak et al., 2010). In the beginning while these travels had the purpose of escaping from war or finding a new home (Hacıoğlu, 2006), today people travel for a multitude of reasons: business, curiosity, religion, culture, education, national culture, recreation and entertainment, sports, health, snobizm, visiting friends and relatives, personal shopping, adventure, seeing new places, searching new places, entertainment, escape, diversity, family, socializing (Nicholson and Pearce, 2001; Usta, 2002; Şahin et. al., 2014, p 111).

In the tourism literature there are several studies that indicate there are multiple motivations for why tourists travel. Lundberg's (1971) research revealed different 18 motives that explain why tourists travel. More than 30 years later, the tourism, travel, leisure, and event tourism literature is replete with articles explaining why people travel or why tourists participate in particular activities, and to understand why people select a specific activity. One of the most important studies in the literature, Cohen (1979) that addressed this issue presented a wide range of motivations from leisure to religion (Kay, 2003, pp 601–602). Therefore, it is important to explain "why people travel?" and "with which purposes tourists travel?".

On the other hand, it seems that nationality is a major discriminant variable in characterizing the distinguished tourist segments. This is confirmed by several other studies, which emphasize that tourist perceptions of a destination or the hospitality businesses may vary according to the tourists' countries of origin (Tsitsiloni et al., 2012, p 14).

Therefore, the understanding of definable segments within the total market is of great importance and the broadest potential factor for segmentation is culture. Kozak (2001, p 391) shows that tourist destinations attract visitors from different cultures and countries. It is not useful to examine the satisfaction level of only one specific group of customers. A comparative analysis between groups is required to better understand the relative importance of the factors that attracted each group to the location.

Existing research on the process of selecting a travel destination, or planning a leisure holiday, has tended to follow the general consumer decision-making model which proposes five key stages. When adapted to the hospitality or travel context, the model commences with the "pre-trip" (1) need recognition, (2) information search and (3) evaluation of alternative destinations, followed by the (4) actual conduct of travel (during trip) and concluding with the (5) post trip evaluation phase such as WOM, etc. (Cox et al., 2009, p 744–745). The travel objectives of tourists can change their travel decisions. For instance, in a study which attempted to determine the travel motives of Canadian ecotourists (Eagles, 1992, p 3), while the usual travel purposes of Canadian tourists is being together as a family to feel at home away from home, visit friends and relatives, warm climate, have fun and be entertained, see the maximum in the time available, lakes and streams, meet people with similar interests, go to places where one feels safe, predictable weather, oceanside, change from a busy job, free to act the way one feels, national and provincial parks, beaches for swimming or training, the purposes of travel for Canadian ecotourists only two travel motivation categories are discussed. The features that are desirable are important to the tourist choice of destination. According to Cai and Li (2009) several earlier studies indicate that rural tourists are most often motivated by opportunities to learn and explore nature or different cultures, participate in outdoor activities, or search for peace and solitude. Travelers decide about many issues directly for international destinations like the holiday type, the means of transport, where they will stay and what they will buy (Choi et al., 1999; Peng, 2013; Musinguzi and Leung, 2010; Şahin et al., 2014, p 110). For example, in a study of French tourists investigating their reasons for choosing to visit Antalya, it was determined that the purposes of the visitors were to learn about the local culture and to inform themselves about the region (Karataş ve Ersoy, 2012, p 98).

Therefore, this study is important as it attempts to determine why tourists choose Turkey rather than some other destination. In other words, the study is important to discriminate between and to classify the purposes of citizens' visit resident abroad and foreign tourists who come to Turkey. Thus, in the study, it is assumed that the purposes of visit of international tourists can be described by the following categories such as sea, sun, sand, hospitality, business, curiosity, culture, religion, health, and sport. But this study limits itself to explore how the two groups can be discriminated based on the variables mentioned before. It does not explore why travelers prefer these variables (purposes).

Multivariate techniques are rarely encountered in quantitative studies (except regression) in the tourism literature and the use of methods such as log linear analysis, canonical correlation, discriminant analysis, cluster analysis, multivariate analysis of variance (MANOVA), or scaling (e.g., multidimensional) is limited. Factor, discriminant, and cluster analysis and scaling are grouped together as they are data reduction techniques. Discriminant analysis, a widely employed marketing tool, has received limited attention only three journals over the 10-year period. This infrequent use is surprising given tourism's extensive borrowings from marketing. The previous observations also apply to cluster analysis; both cluster and discriminant analyses were used four times (2.8%) in the 1984–1985 period and twice during 1986–1987 (Reid and Andereck, 1989, p 23–24).

The studies in the tourism literature, which use discriminant analysis is briefly referred below. In Chen and Hsu (2000, p 414), three forward stepwise linear discriminant analyses were performed to determine the factors influencing tourists' trip planning time frame, budgeted travel costs, and length of trip. Wang et al. (2006) investigated total quality management, market orientation, and hotel performance. They use structural equation modeling (SEM) and discriminant analyses on the research drawing evidence from 588 people. Kılıç et al. (2011) attempted to classify the variables belonging to 30 countries, which include samples from almost all over the world with fuzzy clustering analysis by using the tourism statistics given with nine variables from the year 2007. In addition, after clustering, the researchers applied to the data set belonging to the countries and determined correct classification percentages.

In a different research study which aimed to classify the 30 countries according to their tourism indicators via different cluster analysis methods, correct classification rate for the highest number of clusters is determined as three via average shade statistical, Dunn separation coefficient, and discriminant analysis (Giray, 2013, p 37).

With a different perspective discriminant analysis was used to determine the impacts of each of the service quality dimension on tourists' satisfaction. The dependent variable consisted of the German and Russian tourists' overall rating of the experience, which was divided into three groups: (1) satisfied; (2) undecided; and (3) dissatisfied (Atilgan et al., 2003, p 416).

Moreover, in their recent study (Aktepe and Şahbaz, 2010, pp 83–84) used discriminant analysis to determine how five airlines are perceived by consumers. In a different study which investigate motivations and expectations on rural well-being, tourisms different cluster solutions using k-means clustering were used to find the correct number of segments. Altogether 31 different travel motivations based on earlier literature were measured (Pesonen and Komppula, 2010).

Dündar (2014, p 179) attempted to discover the factors within an instrumental and symbolic framework that are important in distinguishing hotel businesses from their rivals by using one-way ANOVA and discriminant analysis. The study drawing evidence from 121 hotels in Antalya attempted to research what contain the all-inclusive systems and what are the differences between the systems called all-inclusive and others (Üner et al., 2007). At the end of the discriminant analysis, the systems are classified into two groups: all-inclusive and ultra all-inclusive.

As seen from this literature review, discriminant analysis is an analysis method commonly used in tourism research but no studies could be found in the literature with the shape discussed in the current study. This study is relevant to developing countries such as Turkey that have a similar socio-economic situation and a significant number of emigrants. Therefore, the study is important in order to distinguish between the reasons for visiting the country by citizens' resident abroad and foreign tourists within the scope of emigration and developing countries. Thus, it is also important for tourism economics, tourism marketing, destination management experts, and policy makers.

7.3 METHODOLOGY

Let X be an $n \times p$ matrix representing n observations on p variables. The aim of the suggested procedure is to order p variables according to their ability to discriminate observations belonging to r different groups (Fkunaga and Short, 1980).

The goal of LDA is to find a criterion for which projected samples are well separated (Hyun et al., 2003). Specifically, LDA seeks a transformation matrix *W* that maximizes the ratio of the between groups variance-covariance matrix to the within-groups variance-covariance matrix. Initially, we consider a within-groups variance-covariance matrix.

$$W = \sum_{a}^{r} \frac{1}{n_a} \sum_{x \notin S_a} (x - g_{á})(x - g_{á})^t,$$

where r is the number of classes, S_α is a set of data within the α th class, and g_α is the mean of the α th class and $\sum_{a=1}^{r} n_a = n$.

Next between-groups variance-covariance matrix B is defined as $B = \sum_{a=1}^{r} \frac{n_a}{n}(g_a - g)(g_a - g)^t$ where $g = \frac{1}{n}\sum_{i=1}^{n} x_i$ then total variance-covariance matrix can be written as V = W + B.

The projection from a p dimensional space to k dimensional space is accomplished by $y = U^t x$, where U is a pxk matrix with the column u_j. Next, we consider y, the samples transformed from the samples x by the matrix U (Anderson, 1984). The columns of optimal U are the generalized eigenvectors u_i that correspond to the largest eigenvalue $W^{-1}Bu_i = \lambda_i u_i$.

We assign an individual x to group S_α on the basis of which $d(x; g^\alpha)$ for $\alpha = 1, 2, \ldots r$ is smaller where $d(x; g^\alpha)$ is a distance between an individual x and the group mean g^α according to W^{-1} scattering matrix. The stepwise method, we select the q variables at q-th step that maximize the sum, where W_q^{-1} is the within-group variance-covariance matrix obtained from q variables and B_q is the between groups variance-covariance matrix for q variables.

In the stepwise method at the first step, we select the variable that maximizes the quotient of B_1/W_1. At the second step, we use the first variable while selecting a second variable that maximizes $Tr\left[W_2^{-1} B_2\right]$ and so on (Froberg, 1979).

7.4 FINDINGS AND RESULTS

Over all data results belong to years between 2009 and 2018 are shown below. The names and abbreviations of the variables in the analyses are as follows: leisure, entertainment, sports and cultural activities (T); visit relatives and friends (V); education and training (E); health and medical reasons (H); religious and pilgrimage (R); shopping (S); transit passengers (Tr); business and incentive travel (B), and other purposes (O).

In the analysis we used pooled within-group in order to evaluate within-group correlations between foreign tourists and citizens resident abroad. It is about travel preferences (choices) of tourists involved in the research. According to Table 7.1, there is a high within-group correlation (0.819) between V (visit friends and relatives) and H (health and medical reasons). This mean is that when tourists come to Turkey to visit relatives and friends, they also use the medical services. On the other hand, the correlation between E (education and training) and V (visit relatives and friends) has an inverse relationship (-0.766). Also, the correlation score between H (health and medical reasons) and B (business and incentive travel) is very low.

TABLE 7.1 Pooled Matrices Correlation.

	Pooled within-groups matrices correlation							
	T	**V**	**E**	**H**	**R**	**S**	**B**	**O**
T	1.000							
V	.214	1.000						
E	-.121	-.766	1.000					
H	.579	.819	-.766	1.000				
R	-.350	-.772	.736	-.765	1.000			
S	.012	.889	-.723	.678	-.491	1.000		
B	.645	-.226	.295	.059	-.119	-.551	1.000	
O	.093	.764	-.751	.570	-.606	.742	-.127	1.000

When Table 7.2 is examined, it is observed that foreigners have a smaller standard deviation proportionally. The purpose of foreigners' travel is not transitive, regardless of the variable considered.

The greater the eigenvalues, the larger the portion of variance in the dependent variable is explained by the function which has the biggest eigenvalue (values above 40% are generally considered good). Because the dependent variable has two categories, it shows that it will be explained by just one function. The model can explain 97.6% (0.98*0.98 = 0.976) of the relationship between the groups.

TABLE 7.2 Group Statistics for Variables.

	Turkish			Foreigners	
	Mean	**Std. Dev.**		**Mean**	**Std. Dev.**
T	20262189.23	3462323.278	T	19144377.87	3265076.556
V	6373112.62	1125032.597	V	3384238.08	757299.249
E	168843.58	50850.949	E	150552.49	49626.895
H	317278.87	129157.536	H	233760.72	106180.404
R	73686.89	34803.056	R	69619.39	33922.991
S	1155583.91	196280.532	S	1117576.28	189171.141
B	1994841.65	268028.265	B	1769502.20	251849.072
O	1158350.39	107295.000	O	1122212.39	133953.310

TABLE 7.3 Discriminant Function.

Function	Eigenvalue	Variance %	Cumulative %	Canonical correlation
1	41.912[a]	100.0	100.0	.988

First 1 canonical discriminant functions were used in the analysis.

TABLE 7.4 Wilks' Lambda Statistic.

Wilks' Lambda				
Test of function(s)	**Wilks' Lambda**	**Chi-square**	**Df**	**Sig.**
1	.023	62.026	3	.000

Wilks' Lambda tests for every discriminant function the significance of eigenvalue statistical. In the study, significant results are obtained by the significance level % 98.8.

The variables selected for the study have been classified correctly almost 100% in two groups:- foreign tourists and citizens resident abroad. This result shows the success of the analysis.

As a result, discrimination function is obtained as given below:

$$Z = 4.043\ V + 1.808\ R - 2.690\ S$$

When the values including the canonical discriminant function are examined, it is observed that while variables *V*, *R* have positive signs, the other variable *S* has a negative sign. Thus, it has been concluded that these two groups of variables push and pull each other.

Wilks' Lambda (Λ) value is used to test the significance of the discriminant functions and coefficients in discriminant analysis. The significance line (Sig) shows the significance (difference) of the separation power of 3 Factors. In Table 7.7, T as leisure, entertainment, sports and cultural activities; E as education and training; health and medical reasons; Tr as transit passengers; B as business and incentive travel, and O as other purposes. These variables can be removed from the model as these variables do not separate the two groups. At the same time, the first dimension to be analyzed will be the visiting relatives and friends preference, which best distinguishes the group averages with the highest Wilks' Lambda value (Wilks' Lambda 0.270).

TABLE 7.5 Classification.

Classification results[a,c]					
Nationality			**Predicted group membership**		**Total**
			0	**1**	
Original	Count	0	50	0	50
		1	0	50	50
	%	0	100.0	0.0	100.0
		1	0.0	100.0	100.0
Cross-validated[b]	Count	0	50	0	50
		1	0	50	50
	%	0	100.0	0.0	100.0
		1	0.0	100.0	100.0

[a]100.0% of original grouped cases correctly classified.

[b]Cross-validation is done only for those cases in the analysis. In cross-validation, each case is classified by the functions derived from all cases other than that case.

[c]100.0% of cross-validated grouped cases correctly classified.

TABLE 7.6 Standardized Canonical Discriminant Function Coefficients.

	Function
V	4.043
R	1.808
S	-2.690

On the other hand, the variables that are classified by the groups show that, respectively, the variables which have positive value are V (visiting relatives and friends) and R (religious) and the variable which has a negative value is S (shopping). This result indicates that the tourists' purpose of visit to Turkey is visiting relatives and friends, religious, and shopping.

TABLE 7.7 Variables Entered/Removed.

Entered	df1	df2	df3	Exact F			
				Statistic	df1	df2	Sig.
V	1	1	18.000	48.572	1	18.000	.000
S	2	1	18.000	97.814	2	17.000	.000
R	3	1	18.000	223.528	3	16.000	.000

As a result, the standardized discriminant function (Table 7.6) shows the importance of the variables. The factors that best distinguish the groups in terms of their coefficients are V as visit relatives and friends, R as religious and pilgrimage, and S as shopping. At the end of the discriminant analyses, the groups are classified into two groups. They are called as citizen resident abroad and foreign tourists.

7.5 CONCLUSION AND DISCUSSION

The chapter attempts to provide information about specific attributes to be incorporated in the tourism services as per tourists' preferences with the differences between citizens and foreign tourists. The study focuses on the relative importance accorded to the before identified nine salient attributes leisure, entertainment, sports and cultural activities; visit relatives and friends; education and training; health and medical reasons; religious and pilgrimage; shopping; transit passengers; business and incentive travel and other purposes.

Thus, it was determined that these two groups were well separated. In addition, the motivations of these two groups who visit Turkey are revealed. It has been shown that the motivations that significantly separate these two groups of tourists are visiting relatives and friends, shopping and religious and pilgrimage. While visiting relatives and friends is an expected result, religion and shopping stand out from others. Such a finding is important in guiding similar studies. It will also help managers in the sector to determine the direction of the tendency of the groups. The study is important in terms of understanding the preferences of foreign tourists and immigrant citizens coming to developing countries. Moreover, countries with similar structures to Turkey can benefit from the results of study. According to some socioeconomic variables, tourists play an important role in shaping the importance of basic service organizations. Providers of these services should be sensitive to the nature of these variables in the future.

The research has significant implications for tourism management and marketing. The research can enable government tourism agencies with data (about tourist preferences) from the management view, so that they can add importance to their connection with visitors by integrating the desired mix of characteristics. Through the development of corrective action plans, decision-making officials in the tourism department can also evaluate the data supplied by the research to adequately bridge the gaps between their perception of value (of services supplied) and the perception of value by visitors (required services). Such corrective measures will guarantee higher customer satisfaction and differentiable competence.

Consequently, the findings from the discriminant analysis demonstrate that there are significant differences between the two groups. Because those groups indicate that the purpose of travel to Turkey is to visit relatives and friends, religious and pilgrimage and the other important purpose of travel is shopping. The results are very important in terms of actors who are interested in the tourism sector in the scope of macro and micro like economists, marketing and sales specialists, hotel businesses, travel agencies, recreation and entertainment companies, health and medical businesses, hospitals and shopping centers, etc. Especially on a macro scale, tourism policies and plans can be formulated based on this classification in the future. Because as it is seen, people travel to Turkey mostly for entertainment, sport and cultural activities, or health.

KEYWORDS

- **travel**
- **motivation**
- **touristic preferences**
- **resident abroad**
- **foreign tourist**
- **discriminant analysis**

REFERENCES

Altman, E. I. Financial Ratios, Discriminant Analysis and the Prediction of Corporate Bankruptcy. *J. Finance* **1968,** *23*(4), 589–609.

Anderson, T. W. *An Introduction to the Multivariate Statistical Analysis*, 2nd ed.; Wiley: New York, 1984.

Atilgan, E.; Akinci, S.; Aksoy, S. Mapping Service Quality in the Tourism Industry. *Manag. Serv. Qual.* **2003,** *13*(5), 412–422.

Berno, T.; Bricker, K. Sustainable Tourism Development: The Long Road from Theory to Practice. *Int. J. Econ. Develop.* **2001,** *3*(3), 2.

Cai, L.; Li, M. Distance-Segmented Rural Tourists. *J. Travel Tour. Mark.* **2009,** *26*(8), 751–761.

Aktepe, C.; Pars, S. R. Türkiye'nin En Büyük Beş Havayolu İşletmesinin Marka Değeri Unsurları Açısından İncelenmesi ve Ankara İli Uygulaması. *C.Ü. İktisadi ve İdari Bilimler Dergisi*, Cilt 11, Sayı 2, 2010.

Chen, J. S.; Hsu, C. H. C. Measurement of Korean Tourists' Perceived Images of Overseas Destinations. *J. Travel Res.* **2000,** *38*, 411–416.

Choi, W. M.; Chan, A.; Wu, J. A Qualitative and Quantitative Assessment of Hong Kong's Image as a Tourist Destination. *Tour. Manag.* **1999,** *20*(3), 361–365.

Cox, C.; Burgess, S.; Sellitto, C.; Buultjens, J. The Role of User-Generated Content in Tourists' Travel Planning Behavior. *J. Hosp. Mark. Manag.* **2009,** *18*(8), 743–764.

Dimitras, A. I.; Zanakis, S. H.; Zoponudis, C. A. Survey of Business Failure with an Emphasis on Prediction Methods and Industrial Applications. *Eur. J. Oper. Res.* **1996,** *90*, 487–513.

Dündar, A. Örgütsel Çekicilik Bağlamında Otel İşletmelerini Rakiplerinden Ayıran Özellikler. *Anadolu Üniversitesi Sosyal Bilimler Dergisi* **2014,** *14*(1), 175–186.

Eagles, P. F. J. The Travel Motivations of Canadian Ecotourists. *J. Travel Res.* **1992,** *31*(2), 3–7.

Froberg, C. E. *Introduction to Numerical Analysis*; Addison-Wiley, 1979.

Fukunaga, K.; Short, R. D. A Class of Future Extraction Criteria and its relation to the Bayes Risk Estimate. *IEE Trans. Inf. Theory* **1980,** *26*(1).

Giray, S. Ülkelerin Turizm İstatistikleri Bakımından Farklı Kümeleme Analizi Metotları ile Sınıflandırılması ve Türkiye'nin Bu Oluşumdaki Yeri. *International Conference On Eurasian Economies*, St Petersburg, Russia, 2013; pp 695–704.

Hacıoğlu, N. *Seyahat Acentacılığı ve Tur Operatörlüğü*. Nobel Yayın Dağıtım, 2006.

Hyun, C. K.; Daijin, K.; Sung, Y. B. Extensions of LDA by PCA Mixture Model and Class-Wise Features. *J. Pattern Recognit.* **2003,** *36*, 1095–1105.

Karataş, M.; Ersoy, A. Fransız Turistlerin Seyahat Amaçlarının Sosyal, Demografik ve Ekonomik Faktörler Açısından İncelenmesi: Antalya İli Örneği. *EUL Journal of Social Sciences (III:II) LAÜ Sosyal Bilimler Dergisi*, Aralık, 2012; pp 83–101.

Kay, P. Consumer Motivation in a Tourism Context: Continuing the Work of Maslow, Rokeach, Vroom, Deci, Haley and Others, *ANZMAC 2003*, Conference Proceedings Adelaide 1–3 December 2003, Conceptual Papers/Marketing Theory Track, 2003; pp 600–614.

Kılıç, İ.; Emir, O.; Kılıç, G. Bulanık kümeleme analizi ile ülkelerin turizm istatistikleri bakımından sınıflandırılması. *İstatistikçiler Dergisi* **2011,** *4*, 31–38.

Kozak, M. Comparative Assessment of Tourist Satisfaction with Destinations Across Two Nationalities. *Tour. Manag.* **2001,** 22, 391–401.

Kozak, N.; Kozak, M. A.; Kozak, M. *Genel Turizm İlkeler-Kavramlar*. Detay Yayıncılık, (Gözden Geçirilmiş 10. Baskı), Ankara, 2010.

Lundenberg, D. E. Why Tourists Travel. *Cornel. Hotel Restaur. Adm. Q.* **1971,** *26*, 75–81.

Musinguzi, D.; Leung, R. Service Quality Satisfaction of Tourists in Hong Kong. *Fla. Int. Univ. Hosp. Rev.* **2010,** *28*(2), 50–71.

Nicholson, R. E.; Pearce, D. G. Why Do People Attend Events: A Comparative Analysis of Visitor Motivations at Four South Island Events?. *J. Travel Res.* **2001,** *39*(4), 449–460.

Öztürk, M.; Bayat, M. Uluslararası Turizm Hareketlerinde Sağlık Turizminin Rolü ve Kalite Çalışmalarının Önemi Bir Literatür Çalışması. *Kahramanmaraş Sütçü İmam Üniversitesi*, İ.İ.B.F. Dergisi, Sayı: 2, 2011.

Peng, Y. Customer Satisfaction of Mainland Chinese Tourists in Cairns, Australia. *Int. J. Bus. Tour. Appl. Sci.* **2013,** *1*(2), 42–50.

Pesonen, J.; Komppula, R. Rural Wellbeing Tourism—Motivations and Expectations. *J. Hosp. Tour. Manag.* **2010,** *17*(1), 150–157.

Reid, L. J.; Andereck, K. L. Statistical Analyses Use in Tourism Research. *J. Travel Res.* **1989,** *28*, 21–24.

Sheldon, P.; Park, S. Y. *Development of a Sustainable Wellness Destination.* In *Wellness and Tourism. Mind, Body, Spirit, Place*; Bushell, R., Sheldon, P. J., Eds.; Cogniznat Communication: New York, 2009; pp 99–113.

Sussmann, S.; Rashcovsky, C. A Cross-Cultural Analysis of English and French Canadians' Vacation Travel Patterns. *Int. J. Hosp. Manag.* **1997,** *16*(2), 191–208.

Şahin, B.; Sönmez, B; Kahveci, H. Kişilerin Uluslararası Seyahat Engellerini Ölçmeye Yönelik Bir Araştırma. *Gazi Üniversitesi Turizm Fakültesi Dergisi* **2014,** 109–126.

Tsitsiloni, M.; Grigoroudis, E.; Zopounidis, C. Service Quality Evaluation in the Tourism Industry: A SWOT Analysis Approach. *Working Paper 2012.02* **2012.**

Usta, Ö. *Genel Turizm*. Anadolu Matbaacılık, İzmir, 2002.

Üner, M. M.; Sökmen, A.; Güler, D. Her şey dâhil sisteminde farklı uygulamalar ve Antalya ölçeğinde bir araştırma. *Anatolia: Turizm Araştırmaları Dergisi* **2007,** *18*(1), 53–63.

Wang, F. C.; Hung, W. T.; Shang, J. K. Measuring Pure Managerial Efficiency of International Tourist Hotel in Taiwan. *Serv. Ind. J.* **2006,** *26*, 59–71.

CHAPTER 8

THE IMPORTANCE OF ENVIRONMENTAL MANAGEMENT PRACTICES IN HOTELS

U. AVCI[1], U. TURGUT[2*], and F. GÜMÜŞ DÖNMEZ[3]

[1] *Faculty of Economics and Administrative Sciences, Muğla Sıtkı Koçman University, Turkey*

[2] *Graduate School of Social Sciences, Muğla Sıtkı Koçman University, Turkey*

[3] *Faculty of Tourism, Muğla Sıtkı Koçman University, Turkey*

[*] *Corresponding author. E-mail: aumut@mu.edu.tr*

ABSTRACT

This chapter of the book explains the significance of environmental management practices for hotels. In this regard, environmental management practices are defined and the historical development of them is illustrated. Environmental management practices have emerged in order to prevent the negative effects of businesses on the ecological environment in the process of ensuring their sustainability. Discussions about the effects of business activities on ecological environment started after World War II. Also, the negative effect of tourism on the ecological environment has emerged because of mass tourism. Enterprises develop different practices such as energy use, waste management, water conservation, air quality, noise pollution, and visual pollution in order to both protect the ecological environment and assure sustainability. All stakeholders particularly guests, employees and tour operators must be included to the applied by hotels with the aim of reaching success in ecological management practices. To be successful, especially in water conservation and energy use, guests must be involved in the process. In this section, successful practices in waste management, energy use and water conservation are presented as examples.

8.1 INTRODUCTION

Tourism is an activity that majorly depends on the ecological environment (Curtin & Busby, 1999). Pristine beaches in a destination, world resources like clean air and beautiful landscapes accelerate the development of tourism (Mensah, 2019). In fact, such natural resources have become the main attraction centers for various types of tourism across the world (Farrel and Runyan, 1991). However, the increase in mass tourism has caused destruction of the natural resources in many destinations (Gabarda-Mallorqui et al., 2017). This draws the reactions of NGOs, public organizations, and particularly tourists. As a result of these developments, issues like green resort, ecolabel, ecotourism, and sustainability have become a trend in the tourism industry and also made environmental management for hotels more essential than ever (Ouyang et al., 2019).

Hotels not only affect the environment but they are also affected by the environment in the tourism-environment relationship. In this respect, the interaction of hotels with the environment is usually complicated and controversial (Irandu, 2006; Mensah, 2019). Hotels create negative impacts on the natural environment especially by starting from the construction phase. Particularly, owing to the production of luxury products, natural resources such as water and energy are overused in the service production process, and this leads to the rapid depletion of resources (Grove et al., 1996; Mbasera et al., 2016). According to Potrich et al. (2019), the negative impacts of hotels on natural resources after 1990s have made the issue of environment more significant for hotels in comparison to the past. Moreover, the increasing pressure by external stakeholders such as activists has accelerated the adoption of environmental management practices by hotels (Cordano et al., 2010). Within this scope, the aim of environmental management practices for hotels is to detect the environmental damages of hotel activities and implement environment friendly practices (Mutlu and Yıldız, 2015). On the other hand, studies show that very few hotels especially in developing countries have adopted environmental management practices (Turgut, 2013). In recent years, some reasons such as customer demands, the pressure of tour operators, and broad media coverage of environmental issues have been increasing the adoption of environmental management practices by hotels (Chang and Wong, 2006). In this part of the book, as emphasized above, the importance of environmental management practices at hotels is discussed, and the information about their scope is provided.

8.2 DEFINITION OF ENVIRONMENTAL MANAGEMENT PRACTICES

Environmental management is a complex phenomenon including several variables like science, politics, education, and sociocultural factors. Environmental management tries to find solutions for problems that people have while using the natural environment and provide sustainability of the environment. Environmental management practices involve issues such as the fair distribution of resources in the natural environment and protection of them for future generations. In this sense, Barrow (1995) states that environmental management is related to the vital activities of humans because of the fact that human activities have impacts on the environment.

Environmental management practices are policies which are established by businesses to prevent problems arising in the process of sustainability by considering environmental concerns, as well as realization of activities which are planned with respect to these policies (Turgut, 2013). Environmental management practices aim to prevent rapid deterioration of natural resources and its damages created in the process of service provision. According to Hoffman (2000), the purpose of environmental management practices is to provide a balance between the environmental expectations of the business and its stakeholders as a business strategy. Local people, customers, employees, and the state are the primary stakeholders of a business. A successful environmental management practice involves precautions taken to protect the environment from the negative impacts of the industry and make resources sustainable (Meyer, 2000).

Meade and Pringle (2001) think that the fundamental philosophy of environmental management is to do business without damaging the natural environment and disturbing the natural balance. For such business management, there is a need for waste management, energy use, and pollution control practices (Frondel et al., 2018). While environmental management practices contribute to sustainable development on a macro scale, they also offer hotels an opportunity to save money, reduce the costs, and ensure customer and employee satisfaction on a micro scale. According to Starik and Markus (2000), such practices make sure to increase the number of environment friendly people in the long term. Furthermore, Shrivastane and Hart (1995) suggest that environmental management practices provide a route and perspective for businesses to deal with environmental issues. Within this framework, they cover so many organizational issues from input supply and production process to waste management.

8.3 HISTORICAL DEVELOPMENT OF ENVIRONMENTAL MANAGEMENT PRACTICES

Even though destruction of the environment is as old as the history of mankind, debates about the impacts of business activities on the environment have arisen after the Second World War (Shrivastava and Hart, 1994). After the war, environmental issues began to be discussed on a global scale, and the first mass movements emerged in the United States in the year 1968. In this regard, Friends of the Earth developed a policy related to the environment in 1969 and then legalized it (Akdoğan, 2003).

Until 1970s, states had requested from businesses to take the essential precautions in order to reduce and eliminate the damage which they caused on the environment during the production process (Marrow and Rodinelli, 2002). In 1972, representatives of 144 countries, the majority from European countries, gathered at the United Nations Environment Conference in Stockholm arguing that environmental protection measures should be taken in the development process. After the second meeting held in 1977, the third in 1982, and the fourth in 1987, the "World Commission on Environment and Development" was founded. The commission published a report called "Our Common Future" in 1987 (Büyükgüngör, 2006; Hsieh, 2012; Karaer and Pusat, 2002). After this report, the issue of environmental management attracted a great deal of attention on a national and international scale.

Until 1990s, the issue of environmental management had been discussed regarding the industrial sector and factories in general. According to Hutchinson (1996), the negative impacts of the service sector, the silent destroyers of the environment, on the environment have been ignored (Mensah and Blankson, 2013). After the summit in Rio in 1992, practices based on the environmental protection in tourism enterprises, particularly in the hospitality industry accelerated. The International Hotels Environment Initiatives (IHEI) was launched in 1993 by the Prince of Wales after the summit. Within this scope, 11 hotel chains conducted a campaign to improve their environmental performance (Chang and Lam, 2001). In 1994, 16 countries from the Asia Pacific adopted the campaign and started the Asia Pacific Hotels Environment Initiative as the first regional environmental movement (Mackie, 1994). Moreover, in 1994, the Hotel & Catering Institute Management Association joined the Green Globe established by the World Travel Tourism Council aiming at shedding light on environmental management (Anonymous, 1994).

After the significance of environmental management practices toward ensuring sustainability in the hospitality industry was realized by customers and other stakeholders in the 2000s, the issue often remained on the agenda for hotels. At the beginning, the environmental management practices in hotels were seen only as solid waste management. However, within the emergence of standards such as ISO 14000 and 14001 in the following periods, it has involved issues such as energy saving, water usage, visual pollution, and noise pollution. Nowadays, hotels are moving toward building environment friendly facilities from the construction stage of buildings and providing service this way.

8.4 THE IMPORTANCE OF ENVIRONMENTAL MANAGEMENT PRACTICES IN HOTELS

Hotels are enterprises whose integration with the ecological environment is high. According to Rodrigues-Anton et al. (2012), low level of environmental impacts of hotels during the service production process increases the sustainability of their activities. Accordingly, environmental management practices are of great importance for hotels since the aim of environmental management practices is to operate without harming natural resources and the ecological environment (Kirk, 2008). According to Bresciani (2013), the success of environmental management practices in hotels depends on the condition that these practices are a part of the business strategy. However, this condition depends on the size of the enterprise. Additionally, the success in these practices has an impact on customer satisfaction.

The work to do in hotels within the scope of environmental management practices varies in comparison to industrial enterprises. In this respect, hotels should make a difference in (1) environmental consciousness in purchasing, (2) waste minimization and recycling, (3) environmental awareness of customers and staff, (4) social responsibility, and (5) information sharing within the scope of their environmental management practices. Hotels are enterprises which produce services, and customers take part in the production process. Most of the inputs provided by hotels are seen or used by customers. Therefore, not only staff but also customers know whether purchased materials are recycled and environment friendly or not. Hotels generate large amounts and different types (liquid, solid, plastic, etc.) of waste. In terms of environmental sustainability, decomposition and recycling of waste (paper, glass, waste oil, etc.) are important for the success of environmental management practices.

Another aspect that makes environmental management practices distinctive in terms of hotels is the reflection of the environmental awareness of both employees and customers on the enterprise. The success of environmental management practices will ensure satisfaction and even loyalty in terms of customers. In terms of employees, it will increase their satisfaction and commitment to the enterprise. Hotels should achieve successful practices from their business in order to raise awareness in their stakeholders on environmental management practices. To do this, they might create afforestation areas, provide support to educational institutions, introduce green practices to local people, or sponsor them. The final task for hotels is to share their environmental management practices with other enterprises, stakeholders, or researchers (Aslanertik and Özgen, 2007).

8.5 THE DIMENSIONS OF ENVIRONMENTAL MANAGEMENT PRACTICES IN HOTELS

The success of environmental management practices depends to a large extent on the correct implementation of the dimensions of the practices. Thus, first of all, it is necessary to determine the extent to which hotels will focus on environmental practices. Studies on sustainability and environmental management indicate that the practices that hotels will implement on issues such as energy usage, waste management, water usage, air quality, visual and noise pollution will affect their success regarding the environment (Mensah, 2019).

8.5.1 ENERGY USE

The energy use in hotels varies in terms of the size of the hotel, its room and bed capacity, customer profile, location of the hotel, the climate zone, and types of the services provided by the hotel. Energy use has become essential for all hotels to save money with the increase in costs these days. Bohdanowicz (2005) thinks energy consumption in hotels is higher compared to schools and state buildings because the service that is delivered is luxury, and so, this situation makes energy saving more significant. Reducing energy consumption for hotels is one of the biggest steps in terms of environmental management (Bohdanowicz, 2005). In this context, Accor Hotels achieved energy savings at the rate of 5.3% in its all hotel groups in 2015 compared to 2011 thanks to the practices it implemented (Accor Hotels, 2016). Similarly, Hilton provided energy savings at the rate of 14.1% in 2014 in comparison to 2009 (Hilton, 2018), and Marriot International achieved savings at the rate of 13.2% in 2015 in comparison to 2007 (Marriot International, 2016). Furthermore, studies show that energy savings put hotels ahead of the competition by offering a cost advantage (Kannan and Kannan, 2016). Energy savings do not only help reduce costs in hotels but they also indirectly contribute to protection of the environment. In this context, most of the technologies used to save energy in hotels are environment friendly products. On the one hand, they save energy, but on the other hand, they prevent harmful emissions (Chan and Lam, 2002).

Energy saving is a matter of concern not only for staff but also for customers and managers as well. The amount of energy consumed by customers for heating, cooling, and ventilation is between 25% and 50% of the total energy consumed at a hotel. Additionally, the share of a hotel's lighting in the energy that is consumed is between 15% and 20% (Sloan et al., 2009). This shows that customers play a significant role in energy savings. On the other hand, the implications of this are not applicable for industrial enterprises.

The amount of energy consumed to obtain hot water in hotels is another area in the saving process. The energy consumed for hot water changes based on the location of the hotel (the number of sunny days, alternative energy-based heating), but it is usually around 15%. In this context, since resort hotels benefit from solar energy systems, and thermal hotels benefit from hot springs, environmental damage and costs are also reduced. However, Topaloğlu and Gümüş (2012) state in their study about resort hotels that hotels are mostly consume environment hazardous fuels such as fuel oil, coal, LPG, and LNG instead of solar energy systems for heating their water. Hotel managers state that the reasons for this situation are high investment costs and visual pollution of solar energy systems.

The most significant department at hotels where so many energy resources are consumed together is the kitchen. Leakages in kitchen installations, old cooling and heating systems, and absence of filtration systems prevent energy savings. Turgut (2013) determines in his study on 4- and 5-star hotels in Turkey that there is no efficient planning for energy consumption in hotel kitchens. According to Kirk (2008) and Sloan et al. (2009), in order to ensure energy saving in hotels, firstly, every department should make an energy saving plan in itself. Thanks to such planning, how much energy is consumed could be seen more clearly, and in this direction, saving precautions could be taken more easily and in a target-driven manner.

8.5.2 WASTE MANAGEMENT

Hotels are enterprises where consumption is high. Heavy consumption also causes a large amount of waste. Within the historical process, hotels had given negative reactions to especial attempts to decrease solid wastes until 1990s. The main reason for this is the idea that the practice to decrease wastes will increase costs (Wilco et al., 2001). However the other hand, in the upcoming years, it has been realized that solid waste management makes a significant contribution to hotels with regards to both reduction of environmental pollution and providing sustainability (Shamshiry et al., 2011). According to Kirk (2008), there are four main reasons why waste management has gained significance in hotels. These reasons are (1) laws, (2) increase in the number of environmentally conscious customers, (3) increase in environment-oriented touristic activities, and finally (4) financial issues. A successful waste management system offers a saving possibility to hotels in the long term.

Since hotels are consumption-oriented enterprises, they produce more waste than other service businesses. Previous studies have indicated that hotel waste changes based on the characteristics of hotels, but it is about 1 kg per customer (Han et al, 2010). According to Sloan et al. (2009), the 3R (reduce, reuse, and recycle) model is necessary to manage waste in hotels. *Reduce* means reducing or eliminating wastes before they emerge by the correct purchasing policies. This way, avoidance of costs emerging from protection of sources, unnecessary purchasing, storage, transportation, and production processes could be achieved (Kirk, 2008). Hotels are decreasing their wastes by using some methods. Thus, hotels encourage their customers to reuse their towels and sheets instead of changing them daily. This practice helps hotels in terms of water and energy savings as well as for silinecek reduction of usage of chemical cleaning materials (Sloan et al., 2009). The most important amount of waste in hotels is food waste. Especially in all-inclusive systems, 50% of the food and beverages are not consumed, and they form a considerable amount of waste. Chong et al. (2009) state that some hotels in Hong Kong have started to charge their customers for unconsumed foods to solve this problem. *Reuse* means the use of a product for the same or different purposes without changing its original state (Brady, 2005). In hotels, there are so many products and materials which are suitable for reuse. The use of guest towels as rubber, formation of drinking bottles from returnable products, donation of leftovers to animal shelters, or usage of the leftovers in the breakfast for the main dish could be given as examples of reuse. *Recycle* means the process of collecting and processing materials that would otherwise be thrown away as trash and turning it into new products. For this reason, if hotels support recycling and take advantage of this on the maximum level, they could gain an economic profit. To illustrate, recycling of newspapers, papers, brochures, oils, and plastic bottles used in hotels play a significant role in minimizing environmental damage (Kirk, 2008).

8.5.3 WATER CONSERVATION

Hotels are also enterprises where water usage is high, and water is an important cost for hotels. Since popular destinations across the world are generally within warm climate regions, and precipitation in those regions are insufficient, the requirement of water has become much more significant (Turung and Kumar, 2005). Based on estimations, daily water consumption of a customer in a 5-star hotel varies between 170 and 440 l (Han et al., 2010). The large part of water needed for hotels is consumed in the rooms and the kitchen. The study conducted by Kirk (2008) indicates that 37% of the water consumption in hotels is used in the guest rooms, 21% is used in kitchens, 17% is used in public restrooms, 13% is used in other areas, and the remaining 12% is used in laundry rooms. For this reason, the cost of water consumed in hotels is of great importance for hotels. The share of water in the general costs is between 15 and 20% for hotels (depending on the size, types of service).

The amount of water consumption in hotels also increases other energy costs. In this regard, processes like heating or distributing water cause additional costs. Several tips have been developed to determine how efficiency and effectiveness of water consumption in hotels will be increased. Kirk (2008) underlines that it is possible to decrease water consumption in hotels at the rate of 40% based on these tips. It is recommended to pay attention to the tips below to reduce water consumption in hotels (Hsieh 2012; Mbasera et al., 2016; Mensah, 2004).

- Consumption of low-pressure water fittings,
- Determination of water leakages in the water system,
- Accumulation of rain waters for consumption in garden irrigation or toilet cleaning,
- Consumption of treated waters in watering the grass,
- Reuse of towels and sheets,
- Planting that does not require water in accordance with the climate of the hotel.

The Accor Hotels, by adopting the tips above, achieved 8.4% water saving in 2015 compared to 2011 (Accor Hotels, 2016); Hilton achieved savings of 14.1% in 2014 compared to 2009 (Hilton, 2018), and Marriot International achieved savings of 10.4% in 2015 compared to 2007 (Marriot International, 2016). In recent years, it is seen that hotels have made and should make fundamental changes in their facilities to save water to ensure environmental sustainability and combat rising costs (Page, 2009).

8.5.4 AIR QUALITY

Air quality is concerned with the rise of various pollutant factors such as dust, smoke, gas, smell, and water vapor to such a degree that they could damage human beings and other living beings (even objects) (Turgut, 2013). Presumably, 160–200 kg of carbon dioxide gas is released to the air in a hotel per square meter per year (Han et al., 2010). Studies indicate that 213 million tons of pollutants are added to the atmosphere annually as a result of the pollutant factors used by humans. It is stated that 42.6% of them are caused by transportation, 35.21% are caused by industrial and energy facilities, 17.37% are caused by forest fires, and 5.16% are caused by solid waste (Ertürk, 2009).

In city hotels, people spend most of their time indoors. Therefore, especially in city hotels, indoor air quality is important for human health. According to Kirk (2008), indoor air pollution has started to threaten human health in recent years. The increase in the amount of time spent indoors, particularly due to urbanization, has made indoor air cleaning significant (Kirk, 2008). As hotels are places where people come to rest, their air quality is more significant than other enterprises. Additionally, air quality in hotels, a part of the service sector, affects staff health and job performance, as well as customer satisfaction. According to Kuo et al. (2008), low air quality in hotels leads to customer complaints, poor performance in employees, some respiratory disorders, and health problems such as headaches.

Some fundamental regulations are required to improve air quality in hotels. In this sense, firstly, it is necessary to identify the factors deteriorating the air quality and reducing resources that decrease the air quality. According to Garrison (2009), it is recommended to use air filters to improve air quality in hotels and ventilation systems of restaurants and forbid smoking indoors and encourage the use of bicycles instead of motor vehicles in hotels. Furthermore, Samdin et al. (2012) suggest the use of green certificated products instead of air-polluting machines. Kirk (2008) emphasizes that measuring air quality indoors and taking necessary measures are a technical issue in comparison to other environmental practices. In this context, he suggests that hotels should employ skilled staff or get service from consulting firms.

8.5.5 NOISE POLLUTION

Noise is a collection of sounds which have negative psychological and physiological effects on people. Hotels are places where there are so many disturbing stimuli because of a number of reasons such as location, machinery in operation, and sound systems. Kirk (2008) states that customers who come to a hotel with the expectation to relax are likely to experience dissatisfaction when affected by stimuli causing noise for a certain period of time. Additionally, employees are the ones who work with these stimuli, and they are therefore highly affected by them. While customers encounter stimuli at certain times, employees are constantly under the influence of these stimuli (Turgut, 2013). Studies summarize the sources of noise pollution in hotels as follows (Turgut, 2013):

- Noises in laundries,
- Noises in kitchens,
- Noises of compressors, ventilators, and mechanical installations,
- Noises in delivery and disposal areas,
- Noises of furnaces and heaters,
- The sound of music in halls and discos,
- Conversations and music in bars and cocktail halls,
- Public toilets,
- Pools and recreation areas,
- Recreation areas outside of hotels.

It is essential to produce a number of radical solutions to avoid and prevent noise pollution in hotels. In this sense, detecting noise sources correctly, keeping noisy sites away from rooms, ensuring insulation of buildings during construction phases, scheduling activities causing noise and musical entertainment, turning down devices such as telephones, televisions, and radios in guest rooms are among the primary measures (Kirk, 2008).

8.5.6 VISUAL POLLUTION

Visual pollution means the presence of any unwanted sight that may ruin the aesthetic appeal of a specified area. It could reach a point that disturbs people. Visual pollution leads to distraction, fatigue, unwillingness, inefficiency, difficulty in adaptation, irritability, behavior disorders, pessimism, nausea, psychic trauma, and a decrease in the joy of living (Çolakkadıoğlu and Yücel, 2018; Turgut, 2013). The most significant visual pollution for hotels is the architectural structure of buildings. Creating an architectural style which allows employees to have psychological comfort will reduce the effects of visual pollution. The second source of visual pollution in hotels is interior design. In particular, it is suggested that the connection parts of the lobby and other units should be designed in a way that does not strain eyes. Kirk (2008) suggests that hotels should receive support from an interior architect to overcome this problem. The third reason of visual pollution in hotels is inadequate and poor quality of lightning and color mismatch (Bodur and Kucur, 1994).

8.6 CONCLUSION

Hotels play a leading role in contributing to a region's economy. However, if environmental management practices are implemented without establishing an extensive planning during the operations of hotels, their effects on the environment may be quite massive and harmful. As consumers' awareness about the environment and sustainability has increased in recent years, and they look for hotels that are implementing environmental management practices, hotels are obliged to implement these practices during their operations. Environmental management practices in hotels are implemented by focusing on environmental issues such as energy saving, waste management, water conservation, air quality, noise and visual pollution. In this respect, hotels' adoption of environmental management practices ensures the protection of the natural and architectural structures of the regions where hotels are located and enables hotels to make large cost savings.

KEYWORDS

- **environment**
- **environmental management practices**
- **sustainability**
- **hotels**

REFERENCES

Accor Hotels, Planet 21 Research, 2016. http://www.accorhotels.group/en/commitment/sharingour- knowledge/planet-21-research (accessed June 22, 2019).

Akdoğan, A. *Çevreye Duyarlı Yönetim ve İşletmecilik*, Kayseri Ticaret Odası Yayınları: Kayseri, 2003.

Anonymous. HCIMA Backs New Environmental Initiative. *Hospitality* **1994,** *145*, 16–17.

Aslanertik, B. E. ve Özgen, I. Otel İşletmelerinde Çevresel Muhasebe. *İşletme Fakülte Dergisi* **2007,** *8*(2), 163–179.

Barrow, C. J. *Developing the Environment, Problems and Management*; Longman: London, 1995.

Bodur, S.; Kucur, R. Görüntü Kirliliği Üzerine. *Ekoloji Dergisi* **1994,** *12*, 50–51.

Bohdanowicz, P. European Hoteliers' Environmental Attitudes: Greening the Business. *Cornel. Hotel Restaur. Adm. Q.* **2005,** *46*(2), 188–204.

Brady, J. *Environmental Management in Organizations the IEMA Handbook*. Earthscan Publishing: USA, 2005.

Bresciani, S. Managing Environmental Management and Corporate Strategy: Framework and Instrument. *Int. J. Adv. Manag. Sci.* **2013,** *2*(2), 50–59.

Büyükgüngör, H. Çevre Kirliliği ve Çevre Yönetimi. *Toprak İşveren Dergisi* **2006,** *72*, 9–17.

Chang, W. W.; Lam, J. Environmental Accounting of Municipal Solid Waste Originating from Rooms and Restaurants in the Hong Kong Hotel Industry. *J. Hosp. Tour. Res.* **2001,** *25*(4), 371–385.

Chan, W. W.; Lam, J. C. Prediction of Pollutant Emission through Electricity Consumption by the Hotel Industry in Hong Kong. *Int. J. Hosp. Manag.* **2002,** *21*, 381–391.

Chang, E. S. W.; Wong, S. C. K. Motivations for ISO 14001 in the Hotel Industry. *Tour. Manag.* **2006,** *27*, 481–492.

Chong, W.; Wong, K.; Lo, J. Hong Kong Hotels' Sewage: Environmental Cost and Saving Technique. *J. Hosp. Tour. Res.* **2009,** *33*(2), 329–346.

Cordano, M.; Marshall, R. S.; Silverman, M. How do Small and Medium Enterprises go "Green"? A Study of Environmental Management Programs in the US Wine Industry. *J. Bus. Ethics* **2010,** *92*(3), 463–478.

Curtin, S.; Busby, G. Sustainable Destination Development: The Tour Operator Perspective. *Int. J. Tour. Res.* **1999,** *1*(2), 135–147.

Çolakkadıoğlu, D.; Yücel, M. Çukurova Üniversitesi Yerleşkesinde Görüntü Kirliliğine Neden Olan Etkenlerin Belirlenmesi. *Iğdır Üniversitesi Fen Bilimleri Enstitüsü Dergisi* **2018,** *8*(2), 239–246.

Ertürk, H. *Çevre Bilimleri*; Ekin Yayınevi: Bursa, 2009.

Farrell, B. H.; Runyan, D. Ecology and Tourism. *Ann. Tour. Res.* **1991,** *18*(1), 26–40.

Frondel, M.; Kratschell, K.; Zwick, L. Environmental Management Systems: Does Certification Pay? *Econ. Anal. Policy* **2018,** *59*, 14–24.

Gabarda-Mallorquí, A.; Garcia, X.; Ribas, A. Mass Tourism and Water Efficiency in the Hotel Industry: A Case Study. *Int.J. Hosp. Manag.* **2017,** *61*, 82–93.

Garrison, K. Green and Clean. *Fit. Bus. Pro* **2009,** 51–55.

Grove, S.; Fisk, R.; Pickett, G.; Kangun, N. Going Green in the Service Sector. *Euro. J. Mark.* **1996,** *30*(5), 56–66.

Han, H.; Hsu, J.; Sheu, C. Application of the Theory of Planned Behavior to Green Hotel Choice: Testing the Effect of Environmental Friendly Activities. *Tour. Manag.* **2010,** *31*(3), 325–334.

Hilton, Environmental Impact, 2018. https://cr.hilton.com/environment/ (accessed June 23, 2019).

Hoffman, A. J. Integrating Environmental and Social Issues into Corporate Practice. *Environment* **2000,** *42*(5), 22–23.

Hsieh, Y. Hotel Companies' Environmental Policies & Practices: A Content Analysis of Web Pages. *Int. J. Contemp. Hosp. Manag.* **2012,** *24*(1), 97–121.
Hutchinson, C. Integrating Environmental Policy with Business Strategy. *Long Range Plan.* **1996,** *29*(1), 11–23.
Irandu, E. M. Sustainable Tourism Development on Kenya's Coast: A Hospitality Sector View. *Int. J. Tour. Hosp. Res.* **2006,** *17*(2), 189–209.
Kannan, S.; Kannan, S. The Energy Management Strategies for the Hotel Industry in Papua New Guinea. *Asian J. Bus. Manag.* **2016,** *4*(3), 111–116.
Karaer, F.; Pusat, T. ISO 14001 Çevre Yönetim Sistemi Standardının Otomotiv Yan Sanayiine Uygulanması. *Uludağ Üniversitesi Mühendislik-Mimarlık Fakültesi Dergisi* **2002,** *7*(1), 11–20.
Kirk, D. Environmental Management, In *Handbook of Hospitality Operations and IT*; Oxford: Butterword-Heimemann, 2008.
Kuo, N. W.; Chiang, H. C.; Chiang, C. M. Development and Implication of An Integrated Indoor Air Quality audit to an International Hotel Building in Taiwan. *Environ. Monit. Assess.* **2008,** *147*, 133–147.
Mackie, A. Hotels Turning Green. Asian Hotel and Catering Times. **1994,** *16*(7), 19–22.
Marriot International, Environment Principles, **2019**, https://www.marriott.com/Multimedia/PDF/CorporateResponsibility/Marriott_Environment_Principles.pdf (accessed June 23, 2019).
Marrow, D.; Rondinelli, D. Adopting Corporate Environmental Management Systems: Motivations and Results of ISO 14001 and EMAS Certification. *Euro. Manag. J.* **2002,** *20*(2), 159–171.
Mbasera, M.; Plessis, E.; Saayman, M.; Kruger, M. Environmentally-friendly Practices in Hotels. *Acta Commercii* **2016,** *16*(1), a362.
Meade, Bç.; Pringle, J. Environmental Management Systems for Caribbean Hotels and Resorts. *J. Qual. Assur. Hosp. Tour.* **2001,** *2*(3–4), 149–159.
Mensah, I. Environmental Hotel Practices in US Hotels, 2004. https://www.hotel-online.com/News/PR2004_2nd/May04_EnvironmentalPractices.html (accessed June 22, 2019).
Mensah, I. *Environmental Management Concepts and Practices for the Hospitality Industry*; Cambridge Scholars Publishing: UK, 2019.
Mensah, I.; Blankson, E. J. Determinants of Hotels' Environmental Performance: Evidence from the Hotel Industry in Accra, Ghana. *J. Sustain. Tour.* **2013,** *21*(8), 1212–1231.
Meyer, H. The Greening of Corporate America. *J. Bus. Strategy* **2000,** *21*(1), 38–43.
Mutlu, Ç.; Yıldız, M. S. Konaklama işletmelerindeki çevre yönetimi uygulamalarının ve etkilerinin irdelenmesi: Alanya'da faaliyet gösteren 5 ve 4 yıldızlı konaklama işletmelerinde bir araştırma. *Akademik Bakış Dergisi* **2015,** *49*, 107–126.
Ouyang, Z.; Wei, W.; Chi, C. G. Environment Management in the Hotel Industry: Does Institutional Environment Matter?, *Int. J. Hosp. Manag.* **2019,** *77*, 353–364.
Page, J. P. *Tourism Management: Managing for Change*; Elsevier Ltd.: London, 2009.
Potrich, L.; Cortimiglia, M. N.; Medeiros, J. F. A Systematic Literature Review on Firm-Level Proactive Environmental Management. *J. Environ. Manag.* **2019,** *243*, 273–286.
Rodriguez-Anton, J. M.; Alonso-Almeida, M.; Celemin, M. S.; Rubio, L. Use of Different Sustainability Management Systems in the Hospitality Industry. The Case of Spanish Hotels. *J. Clean. Prod.* **2012,** *22*, 76–84.

Samdin, Z.; Bakori, K. A.; Hassan, H. Factors Influencing Environmental Management Practices Among Hotels in Malaysia. *Int. J. Econ. Manag. Sci.* **2012,** *6*, 84–87.

Shamshiry, E.; Nadi, B.; BinMokhtar, M.; Komoo, I.; Saadiahhashim, H.; Yahaya, N. Integrated Models for Solid Waste Management in Tourism Regions: Langkawi Island, Malaysia. *J. Environ. Public Health* **2011,** 5.

Shrivastava, P.; Hart, S. Creating Sustainable Corporations. *Bus. Strategy Environ.* **1995,** *4*, 154–165.

Shrivastava, P.; Hart, S. Greening Organizations 2000. *Int. J. Public Adm.* **1994,** *17*(3–4), 607–635.

Sloan, P.; Legrand, W.; Chen, J. S. *Sustainability in the Hospitality Industry Principles of Sustainable Operations*; Elsevier Linacre House, Oxford: UK, 2009.

Starik, M.; Marcus, A. Introduction to the Special Research Forum on the Management of Organizations in the Natural Environment: A Field Emerging from Multiple Paths, with Many Challenges Ahead. *Acad. Manag. J.* **2000,** *43*, 539–547.

Topaloğlu, C.; Gümüş, F. Yenilenebilir enerji kaynaklarının konaklama işletmelerinde kullanımı: Muğla İli'ndeki 4 ve 5 yıldızlı konaklama işletmelerinde bir araştırma, 13. Ulusal Turizm Kongresi/ Akdeniz Üniversitesi, 6–9 Aralık 2012, 1039–1050.

Trung, D. N.; Kumar, S. Resource Use and Waste Management in Vietnam Hotel Industry. *J. Clean. Prod.* **2005,** *13*, 109–116.

Turgut, U. Kıyı Otellerinde Çevresel Yönetim Uygulamaları. *Yayınlanmamış Yüksek Lisans Tezi*, Muğla Sıtkı Koçman Üniversitesi, 2013.

Wilco, W.; Lam, C. J. Environmental Accounting of Municipal Solid Waste Originating from Rooms and Restaurants in the Hong Kong Hotel Industry. *J. Hosp. Tour. Res.* **2001,** *25*, 371.

CHAPTER 9

A CRITICAL EVALUATION OF SOCIAL AND ENVIRONMENTAL IMPACTS OF TOURISM DEVELOPMENT IN TURKEY

M. TUNA[1] and M. MUTLU[2*]

[1]*Department of Sociology, Faculty of Literature, Muğla Sıtkı Koçman University, Muğla, Turkey*

[2]*Department of English Teaching, Faculty of Education, Muğla Sıtkı Koçman University, Muğla, Turkey*

[*]*Corresponding author. E-mail: mtuna@mu.edu.tr*

ABSTRACT

Tourism and its effects have been seen primarily in the Southern Aegean and Western Mediterranean regions of Turkey. The faster social, economic and environmental changes related to tourism are particularly important in the coastal cities of Turkey. Antalya, Muğla, Aydın and İzmir provinces are providing important examples. Significant socio-economic changes have appeared in these areas during the last 30–40 years as a result of tourism development. Before that time, the above-mentioned provinces were small agricultural and fisheries towns. It is clear that direct and indirect factors of general economy, such as global economic fluctuations, supply and demand factors and structural problems of tourism in Turkey, have affected the social and economic structures of these cities.

The impacts of tourism development are multi dimensional and include economic, social and environmental impacts. Since there is already abundant research on the economic impacts of tourism development, this study will focus more on the social and the environmental impacts as they relate to Turkey; it will also offer some suggestions towards improving these impacts.

9.1 INTRODUCTION

Enormous social and economic changes have been seen in Turkey for the last few decades. Industrial sectors such as automotive, steel industry, power generation, machinery, petrochemistry, and construction sectors presented a huge development for the last 50 years. However, one of the most significant social and economic changes has been seen in service sector. Namely, tourism is one of the fastest developed economic sectors in Turkey. The globalization and internationalization of tourism trends effected this fast-growing tourism development in Turkey. Therefore, Turkey's tourism structure has been changed from national level to international level, and tens of millions of international tourists have visited Turkey and billions of USD income has been generated in the tourism sector since the 1990s. More than 40 million international tourists and approximately 40 billion USD income were expected in the year 2019. Therefore, tremendous environmental and social changes have been witnessed as a result of this swift development in industrial sectors and tourism sector for the last 50–60 years in Turkey. Main social changes have been witnessed in the social structure, and social structure has changed from traditional form to modern form. Additionally, environmental degradation, environmental pollution in the air, water, and soil, desertification, diminishing of agricultural lands has been seen; especially in the industrial areas and coastal tourism areas. Accordingly, the social and environmental impacts of swift economic development, especially impacts of tourism development on society and environment, are investigated in this chapter. Moreover, a comprehensive and critical perspective is rendered in this evaluation. This chapter also comprehends a detailed review between society and environment relationship.

Turkey as a developing country that presented a swift social and economic development since World War II. The economic indicators of the mentioned period presented that national income and per capita income enormously increased and social welfare increased too. On the other hand, so many negative environmental impacts such as pollution in the air, water, and soil, desertification, diminishing of agricultural lands have appeared in this period. Accordingly, environmental interests and environmental perception of Turkish society have been changed and society has become much more sensitive to environmental issues. In view of fact that if we look at the environmental perception of Turkish society, we have seen that Turkish society is in between demand of socioeconomic development and environmental sensitivity. On the one hand, Turkey as a developing country demands socioeconomic development and welfare; on the other hand, the natural environment needs a sensitive and sustainable usage of natural resources and natural environment. However, Turkish society is most likely on behalf of developmental perspective instead of environmentalist perspective.

The fact that matter affects the social and environmental cost of this swift socioeconomic change is also enormous. As a result of faster industrialization, urbanization, and especially faster tourism development gave a huge negative impact on the natural environment in Turkey. Therefore, mainly environmental impacts of faster socioeconomic development; especially impacts of industrial sectors and impacts of tourism development have been evaluated in this chapter. Accordingly, this chapter includes two main subjects: first, subject is a general view of society environment relationship, and second, subject is environmental and social impacts of faster tourism development in Turkey.

First subject includes general view of environmentalism, historical background, and environmental management system in Turkey. Second subject includes definition and conception of tourism, a general view of the social and environmental dimensions of tourism development and social and environmental impacts of tourism development in Turkey. The author of this chapter has written some papers on society–environment relationships in Turkey. The papers are *Public Environmental Attitudes in Turkey*; *Social, Environmental Impacts of Tourism Development in Turkey*; and *Society, Environment and Tourism Relationship in Turkey.* This chapter is considerably compiled and developed from these three papers.

9.2 SOCIETY AND ENVIRONMENT

Environmental issues have been taking more public attention for the last few decades in Turkey and all around the world. Environmental degradation, pollution in the air, water, and soil, diminishing of rural, and agricultural areas, desertification global warming are some of the global environmental issues that impact the whole world. Turkey is also under impact of those global environmental problems; therefore, the emerging environmental issues are one of the most important social problems that affect the society.

Environmental problems and environmental issues have become one of the most important problems and issues in the late modernization age. Earlier, environmental problems have been accounted as technical and economic problems; however, these problems have been accounted as their societal dimensions as well as technical and economical dimensions since the 1950s.

There are two main-stream perspectives to evaluate the relationship between the society and the environment in the environmental sociology literature: the anthropocentric worldview and the eco-centric worldview. The anthropocentric worldview mainly underlines that human being is the dominator of the natural environment and natural environment has valued for human usage. On the other hand, according to the eco-centric worldview, natural environment has valued for not (only) human usage, it is valued for its own sake too. In general, it is assumed that public environmental awareness varies from the anthropocentric form to the eco-centric form. The anthropocentric form of public awareness is more development oriented and definitive human oriented, and the eco-centric form of public awareness orients more environments. Turkish people's environmental attitudes are in between of the anthropocentric form and the eco-centric form of public awareness.

Environmental issues are increasingly taking more public attention in Turkey for the last few decades. Air, water, and soil pollution, desertification, lake, sea water, and river pollution are major environmental issues and problems that heavily affect the Turkish society. As a result of these major environmental problems, environmental issues take more and more public consideration in Turkish society. Therefore, society–environment relationship in Turkey is also investigated in this chapter. The investigation is most likely based upon an overview of public environmental attitudes or environmental values in Turkey.

Before discussion of society–environment relationship in Turkey, a historical and theoretical evaluation of society–environment relationship is accomplished.

9.3 HISTORY OF SOCIETY–ENVIRONMENT RELATIONSHIPS

There is a close association between socioeconomic reproduction type and society–environment relationships. The history of society–environment relationships shows three stages of these relationships. These stages are hunter-gatherers societies, agricultural societies, and industrial societies. Charles Harper (Harper, 1996) investigated historical and social background of society–environment relationships and explained this relationship using paradigmatic model. According to Harper, "a cognized environment is part of the cultural worldview—the totality of cultural belief systems about the world and reality that people share. A cognized environment is also an important component of the paradigms that people share.... A social paradigm is an implicit model of how the world works that is broadly shared by people in society." A social paradigm affects society's institutional structure and moral values. Every society has competing paradigms with one tending to dominate. Although the dominant social paradigm (DSP) is not necessarily the only paradigm in a society, it is supported by the elite of social systems and provides an effective worldview (Harper, 1996; Milbrath, 1984). The DSP culturally and socially legitimates society's worldview by defining social structure and culture.

As Lenski (Lenski, 1966) and Harper noted, each society creates exclusive power structures, division of labor, exploitation and surplus structures, and environmental relations. Historically, each society has created an exclusive relationship with the environment and a DSP that legitimizes the relationship between society and the environment. The relationship of each society with the environment, DSP of these societies, and power structures is discussed in a historical context. Three main types of societies and their relationships with environment are given a historical perspective.

Hunters-gatherers were the earliest type of human society appearing approximately 40,000 years ago (Harper, 1996). Hunter-gatherers lived as small bands, survived by gathering edible wild plants and killing animals from their close surroundings. They collected their foods on daily or weekly basis, so they could not accumulate long-run economic or food surplus. "They survived by the accumulation of a cultural stock of 'expert knowledge'" (Lenski, 1966). The "expert knowledge" includes knowing food availability of their close surroundings, including the knowledge about soil, water, air, and plants. Social and organizational structures of hunters-gatherers were not so complicated. They consisted of a simple division of labor and their status-role system was based primarily on age and gender. Relationships between members of the society were face-to-face and informal, and there was at its lowest level inequality because there was a very little surplus value accumulated (Lenski, 1966).

The relationship between hunter-gatherers societies and the natural environment was a direct relationship compared to modern societies. Due to the smallness of societal population and local and decentralized social patterns, impact of hunter-gatherers on environment was characteristically limited and localized. Hunter-gatherer societies developed a strong association with the nature which nurtured a dependence on nature and other societal members. This dependence on nature was at the core of hunter-gatherers' DSP. They thought of themselves as "as people in nature." "Their cognized environment was that of a living natural world (wilderness/jungle/forest/grassland) of things and beings governed by spiritual forces" (Harper, 1996).

Human beings discovered cultivation of crops and domestication of animal about 10,000 years ago. Irrigation, fertilization, and organization of labor accelerated agricultural production in later stages of agricultural societies. Innovation of more advanced agricultural technologies such as the metal plow, pulled by domesticated animal, enormously increased agricultural production, and radically changed social life in agricultural societies. Division of labor became more significant in social life and a peasantry appeared as the main force of production. Additionally, artisans appeared as creators of agricultural and other tools, and administrative class appeared as organizer of production and surplus. However, 90% of the population was peasant in an agricultural society (Harper, 1996).

The basic difference between hunter-gatherers societies and agricultural societies was inequality and manipulation (exploitation) of natural environment. Leading classes (Lords and landed aristocracy) accumulated and dominated surplus that was produced by peasantry and artisans. This accumulation created a huge inequality between leading classes and the forces of production. Manipulation, overuse of natural resources, and exploitation of nature for agricultural production were the main environmental premises of agricultural society. Moreover, first urban settlements of human history were established in agricultural society. Urban settlers of agricultural society seem to be "uprooted" from their cultural and natural "roots" compared to hunter-gatherer's holistic lifestyle in the natural environment. There was no urban infrastructure such as water, electricity, road, sewage system, education, and public health services in agricultural society, so daily life was misery for poor settlers in the cities.

The social and environmental manipulation and exploitation were not only responsible for the rise of agricultural society; they were reasons for its collapse as well. Socioeconomic exploitation of the peasantry resulted in uprising the elites while environmental overuse and exploitation reduced agricultural production. For example, Harper noted the collapse of Lower Mayan societies in Central America. Lower Mayan societies cleared forests for agricultural land; nevertheless, they lost their cultivated lands by erosion because of the lack of forests. Losing cultivated land resulted with hunger and total collapse of the society.

Industrialization started about 300 years ago in the Western Europe. Industrialization was based upon some key innovations including the development of the textile industry in England. Discovery of the steam engine, electric power, hydroelectric power, and petroleum were also important contributors to industrialization. Development of these new energy sources and production technologies enormously increased industrial production and surplus. Invention of new technologies and increase in production required more centralized social organizations. The centralization and reorganization of production radically changed people's daily life. People accumulated in industrial centers, and eventually majority of the population began to live in urban industrial centers for the first time in human history. Increased urbanization was accompanied by a significant expansion in political and economic bureaucratization (Wallerstein, 1979).

The Western industrial nations established industrial empires. These industrial Western countries imported raw materials and natural resources from the undeveloped countries, and they reexported these materials as manufactured goods to these countries. Expansion of industrialization to an international level also expanded environmental exploitation. Industrialization also increased domination of humans over other species. Clearing exotic species and natural environments was acceptable to establish industrial plants and to increase industrial production.

The environmental impacts of industrialization are huge. Pollution in the air, water, and soil reached the maximum level in the industrialization age. Global warming and global climate change, desertification, water shortage, food shortage, hunger, and nuclear risk are just some of environmental impacts of industrialization in the modern industrial age. The global warming and global climate change are possibly the most effective environmental impacts that ever have been seen.

Harper defined industrial society's DSP as follows: "... if the main cognized environment of agricultural societies was, that of garden to be tended, modified, and dominated by humans, that of industrial societies is a dramatic extension of this." The European Enlightenment prepared cultural (scientific, moral, and philosophical) bases of exploitation and manipulation of nature. After this short historical review, conceptual framing and clarification about environmentalism and environmental attitudes will be presented.

Environmentalism has a cumulative meaning that includes environmental action, environmental movements, environmental politics, and environmental attitudes. Harper indicates that environmentalism is both ideology and action: "as ideology, it is a broad set of beliefs about the desirability and possibility of changing the human relationship with the environment." According to this definition, environmentalism includes environmental beliefs and attitudes. Environmentalism is a total perception or worldview within a society on environmental relations.

Historically, the meaning of environmentalism as a social movement and political ideology varies from the anthropocentric pole to the eco-centric pole. Resource conservation, human welfare ecology, preservation, animal liberation, and eco-centrism are major environmentalist streams (Eckersley, 1992). The resource conservation movement basically advocates conservation of the natural resources for better usage. Human welfare ecology targets a clean, safe, and more pleasing natural environment for human beings. The preservation movement advocates reservation of wilderness for human enjoyment. Accordingly, this movement, like the resource conservation movement and the human welfare ecology movement, is human centric as well. These three environmental streams are anthropocentric streams. According to these streams, environment has instrumental value for us. These environmental streams are confirmed by the DSP of industrial society. However, according to modern environmentalism, nature should have been valued for its own sake (Eckersley, 1992).

Accordingly, environmentalism more likely represents an ecological or eco-centric meaning (Eckersley, 1992). Environmentalism with its ecological meaning also identified as "deep ecology." The term "deep ecology" was first used by Norwegian philosopher Arne Naess. The deep ecology approach, like eco-centrism, identifies a holistic life within the environment. The deep ecology also criticizes "shallow" environmentalism which is also anthropocentric. Shallow ecology movement identifies an uncritical environmentalism position. Shallow ecology movement is against pollution and resource depletion. The main objective of this movement is to make better health and affluence conditions of people in the developed countries (Naess, 1991).

9.4 A GENERAL VIEW OF SOCIETY ENVIRONMENT RELATIONSHIPS IN TURKEY

As it was mentioned earlier, a faster social and economic development has been witnessed in Turkey for the last few decades. The main dimensions of this development are in the industrial sector, such as machinery automotive, steel industry, petroleum, and chemical industry, construction technologies; service sector such as tourism, education, finance, health, and so on. The societal dimension of this faster development has been seen in the social structure of the Turkish society. The social structure of the Turkish society has been changed from more traditional social structure to more modern social structure in abovementioned period. The economic indicators of the mentioned period presented that national income and per capita income increased enormously and social welfare increased too.

However, an enormous environmental change and a huge environmental impact have been seen in the mentioned period. Environmental degradation, environmental pollution in the air, water, and soil, lake and water pollution, desertification, diminishing of agricultural and rural areas have been seen especially in the industrial areas and coastal tourism areas. Therefore, environmental issues and environmental problems are the main components of the swift in socioeconomic development since the World War II in Turkey. Accordingly, environmental interests such as environmental perception, or environmental attitudes of the Turkish society are also taking social scientific interests for the last few decades. If we look at the environmental perception of Turkish society, we have seen that Turkish society is in between demand of socioeconomic development and environmental sensitivity. On the one hand, Turkey as a developing country demands socioeconomic development and welfare; on the other hand, the natural environment needs a sensitive and sustainable usage of natural resources and natural environment. However, Turkish society is most likely on behalf of the developmental perspective instead of the environmentalist perspective.

A general overview of Turkish public environmental tendencies presents that environmentalism level of Turkish society is not high. However, environmental sensitivity of Turkish population is increasing for the last few decades (Özdemir, 1998). Additionally, it might be mentioned that there is an environmentalist movement in Turkey, even if it is not so powerful and effective. Moreover, environmental sensitivity is a relatively new issue in Turkey. The main priorities of Turkish population are socioeconomic development and social welfare, instead of the environmental sensitivity. Therefore, environmental issues are not priority in this context. The main difference between the Western societies' and Turkish society's environmentalism structure is related to their industrialization processes. The Western countries' industrialization processes started in the late 18th century and most likely completed in the early 20th century. However, the non-Western countries industrialization started in the middle of the 20th century. Environmental issues were not really serious problems while the Western countries have been industrialized and they used natural resources abundantly. However, todays' developing countries' industrialization processes are totally different from the Western counterparts. Nowadays, developing countries are using natural resources that are much more expensive and limited. Additionally, there are much more environmental problems and environmental risk for today's word such as air, water, and soil pollution; environmental degradation; global warming; and so on. All these environmental conditions limit developing countries' socioeconomic development levels. Turkey is also a developing country that is affected by abovementioned conditions and it is behalf of the developmental perspective instead of the environmental perspective.

If we look at the environmental position of Turkey within real examples; we should mention so many examples that are related to the environmental position of the country. Especially, electricity generation, mining, and other industry-related issues that are related to the environmental contradiction in Turkey. Coal-burned power generation plants are some of the problem areas of environment-development contradiction. Turkey needs more electricity for industry and social welfare and coal is the main and relatively cheap natural resource to generate electricity power. However, using coal to generate electricity gives enormous polluted chemicals in the air and causes air pollution, acid rain and even greenhouse effect, and global warming. Therefore, using coal for electricity generation is a big discussion point in Turkey that defines environment and development contradiction. The society is in a contradiction between the electricity generation and protection of the natural environment. However, society is on behalf of the electricity generation instead of the protection of the natural environment. Yatağan Thermic Power Plant is one of the examples of this issue as tens of similar examples still exist (Tuna, 2001).

Another big issue that relates environment and development contradiction is mining, especially gold mining. Gold is an expensive exchange material and one of the most important commodities for all history. However, gold mining gives too many polluted materials to the soil. The contradiction point of gold mining emerges in the production process. On the one hand, gold is a very expensive exchange commodity; however, production process of gold releases a lot of polluted chemicals to the natural environment, especially soil and deep water. These chemicals pollute agricultural soil and drinking deep water. Therefore, the society is also in a contradiction in gold mining case and it is on behalf of gold mining instead of the environmental protection. Bergama Gold Mining case is a well-known example that exemplifies the gold mining contradiction. Lately, another popular gold mining discussion is about Çanakkale, Kazdağları. Kazdağları is one of the most well-preserved forest areas with wealthy oxygen deposit. However, a Canadian gold mining company planning a gold mining plant in the area and very strong environmentalist resistance movement appeared against the gold mining operation in the late 2019.

One of the indicators of the society–environment relationship and the environmentalism is environmental values. Environmental values of Turkish society have been investigated by many researchers (Kasapoğlu and ve Ecevit, 2002). One of these researches is much more comprehensive and the research population includes all Turkish population. The research includes environmental issues such as environmental politics, environmental knowledge, environmental degradation, environmental pollution, energy usage, and so on (Tuna, 2004). The results of this research indicate that Turkish society is most likely behalf of basic form of environmentalism (environmental worldview) and less likely behalf of more specific form of environmentalism (environmental commitment). In other words, the society is in between of environmental sensitivity and economic development interests.

9.5 TOURISM, ENVIRONMENT, AND SOCIETY

The tourism development is one of the main factors that affects social and environmental structure of Turkey for the last few decades. The history of tourism development goes back to 1960s in Turkey. As a result of industrialization and urbanization, not only the upper classes, but the middle classes also have taken summer and winter vacations after the 1950s. However, much improvement has not been seen in terms of tourism development in Turkey until 1980s. The last decade of the 20th century is a real milestone in terms of international tourism development in Turkey. A faster international tourism development has been seen since 1990s in Turkey. The main reason for this development is globalization, improvement of communication, and transportation technologies.

Tourism and its effects are primarily seen in the Southern Aegean and Western Mediterranean regions of Turkey. The faster social, economic, and environmental changes related to the tourism development in the coastal towns and cities of Turkey are particularly important. These changes primarily have been seen in the Southern Aegean and Western Mediterranean regions namely coastal towns of Antalya, Muğla, Aydın, and İzmir provinces, providing important examples. Significant socio–economic changes have appeared in these areas during the last 30–40 years as a result of tourism development. Before that time, the above-mentioned towns were small agricultural and fisheries towns. However, these towns have become famous international tourism destinations in the last few decades. As a result of this faster tourism development, so many social and environmental impacts have been seen in these towns. The impacts of tourism development are multidimensional and include economic, social, and environmental impacts. Since there is already abundant research on the economic impacts of tourism development, this chapter will focus more on the social and the environmental impacts as they relate to Turkey; it will also offer some suggestions toward improving these impacts.

Tourism is one of the most well-known free-time and recreational activities in modern society. It is a temporary displacement of people for health, sport, entertainment, travel, and holiday (Lanquar, 1999). A geographical mobility is therefore operative, making travel an important part of tourism. But an operational definition of tourism places less emphasis on travel and more emphasis on accommodation and hospitality. So, even though tourism today is generally considered as free-time and recreational activity, the wider perspective is that traveling for health, business, sport, and scientific congress and meetings also fit the broader meaning of tourism.

All components of tourism should be considered to gain the fullest picture of the concept. When looked at from the economic perspective, for example, tourism is a *service sector* that has production and consumption dimensions. In other words, there are "producers" and "consumers" of the tourism service. The "producers" are the managers, owners, and workers within the tourism sector; the "consumers" are the tourists. There is an economic exchange between the consumers (tourists) and the producers (workers within the tourism sector). Or, when viewed from a sociological perspective, tourism is a *social interaction* between the host community and the guest community (Apostolopoulos, 1996). It is the contention of this chapter that the societal dimension of tourism deserves at least as much consideration as the economic dimension. Therefore, the societal and, along with it, the environmental dimensions of tourism development in Turkey are the main focus of this chapter.

Although normally discussed primarily in economic terms, it should be clear that an understanding of tourism based solely on the economic point of view is insufficient (and ultimately uneconomical). Behind its economic dimension, tourism is a social interaction in which people who otherwise would most likely never meet come together in personal face-to-face interaction (Cohen, 1996). The quality of those interactions is, therefore, crucial and impacts the economic dimension, whether the interactions are between two individuals, two groups, two nations, or two different cultures. These interactions, therefore, also have both national and international tourism implications.

If we look at tourism from the group or societal level, two different groups of people, society, nation, a wider perspective even two different cultures come together and engage in interaction especially in terms of international tourism. Two groups of people engage in a social interaction that most likely their language, religion, and cultures are different. The sociocultural backgrounds of the tourist and the host determine the structural inside of the interaction. The quality of interaction between the tourist and the host affects the economical dimension of tourism. If we look at the tourism activity from an economic perspective, two sides of the economical exchange—the producer and the consumer—should be satisfied (Swarbrooke and Horner, 2001; Baudrillard, 1998). On the other hand, if it has been looked at the tourism activity from a societal perspective, two sides of the social interaction—the host and the guest—should be satisfied within the social interaction. Accordingly, a proper tourism activity is depending upon a healthy economic activity. Then a healthy economic activity is depending upon a proper social interaction between the consumer or tourist community and the producer or host community.

The societal dimension of tourism has been discussed so far; then, a short initial discussion will be given on the environmental dimension of tourism which is the second major subject of this chapter. Several different perspectives are possible to evaluate environmental dimension of tourism. The most radical perspective is possibly the ecotourism perspective which implies a totally different viewpoint about the tourism activity. According to this perspective, the tourism activity or the tourism development should totally be planned within the ecological principles. Another perspective about the environmental dimension of the tourism activity is about the developed tourism areas. The major point of abovementioned perspective is that tourism should be changed in a more environmentalist way or environmentally sensitive way in the developed tourism areas (Tourism and the Environment in Europe, 1995).

9.6 DIMENSIONS OF TOURISM

As it has been mentioned earlier, tourism is not only an economic activity but also has social and environmental dimensions as well; therefore, tourism should be accounted as a multidimensional phenomenon. Tourism is accounted as a service sector in the general economy and affected so many factors. These factors affect the quality of service and productivity of the sector and finally the income generated by the sector. The demands and expectations of tourists changed, and the form and structure of tourism has been changed as well for the last few decades. Therefore, natural, environmental, historical, cultural, artistic, and social factors have become unavoidable factors for new tourism trends (Judge, 1992). The natural environmental factors, cultural and historical values of the destination areas are the major preferences for tourists. Therefore, concepts of preservation of the natural environment, the environmentally friendly tourism products, touristic facilities within the natural environment, and the sustainability come on stage for the new tourism trends (Mowforth and Munt, 1998). Ecotourism is the first concept for a sustainable tourism development. The concept of ecological tourism means that ecological principles should be accounted as the main principles of tourism development. In other words, ecological principles should be considered as main values and the natural resources used within a sustainable way for all stages of tourism development. The principles of ecological tourism are not only considered for new tourism areas, but also they should be accorded with traditional tourism areas and facilities. The European Union as a main tourism market of Turkey has a tendency that the ecological principles should be the main principles of tourism development. The EU Regional Committee was held on November 21, 2001, Florence, Italy, and it was discussed that all rules and regulations related to the tourism development should be accorded with sustainability and ecological tourism principles (Inforegio/Panorama, 2002/2006).

Turkey should consider EU's tendencies about tourism development for two reasons. First, the EU countries are the main tourism market of Turkey and, second, Turkey is an associate member of the EU. Therefore, changing Turkey's tourism structure to more ecological and sustainable way appeared as an unavoidable necessity. Accordingly, major structure and infrastructure of tourism must be changed within the process of the EU membership in Turkey. Therefore, an evaluation of social and environmental impacts of tourism development in Turkey—particularly the Southern Aegean and the Western Mediterranean regions—is an important necessity to understand the social dimension of tourism in Turkey. This evaluation also provides an important understanding of possible tendencies about the future of tourism development in Turkey.

9.7 THE SOCIAL DIMENSION OF TOURISM

If tourism is evaluated from a wider perspective, it has several social impacts as well. Accordingly, the social changes as a result of tourism development grouped as following two major points:

1. *Demographic impacts*: A general increase and changing structure of population in term of age, gender, and education in a given tourism area.
2. *Economic impacts*: Changes in production and consumption forms.
   a) *Changes in consumption forms*: The consumption forms are changing from the traditional, local forms to the national and international forms. An emergence of consumption forms that is popular in the global market.

b) *Changes in production factors*: Changing human resources using and demanding new technologies; educational, occupational, and linguistic changes.

3. *Cultural impacts*: The cultural production and consumption forms of the society have been changed. The traditional life forms and customs have been changed to more modern forms.
   a) *Changes in norms*: Traditional social tests and preferences have been changed. The social behavior that was not accepted by society before the tourism development; however, it has been accepted after that.
   b) *Consumption of the culture*: Entertainment, religious rituals, marriage ceremonies, local arts, and literature that are framed by the traditional culture and custom have been changed to more national and international forms.
4. *Impacts on the natural environment*: A degradation of the ecological, historical, and geographical resources, and fauna and flora. General pollution in the natural environment, particularly pollution on air, soil, and water).

The demographic, economic, and cultural impacts of tourism development are identified as social environment. Then, the impacts of tourism on the environment can be divided as two subjects: impacts of tourism on the natural environment and impacts of tourism on the social environment. Insofar, the social environment is just nested in the natural environment; in other words, the natural environment comprehends the social environment. Therefore, the policies to minimize the negative impacts of tourism development on the environment give a priority to a specific protection of the natural environment.

9.8 THE ENVIRONMENTAL DIMENSION OF TOURISM

The environment might be defined as the natural and artificial factors that are determining human life. The environment has multidimensional factors. These factors are biological, physical, and sociocultural categories. Human beings have always tried to maximize usage of natural resources throughout history. Therefore, human societies created an artificial environment on the natural environment. If it has been looked at from a historical perspective;

the usage and manipulation of the natural environment was very limited in the hunting and gathering societies and the agricultural societies. However, a maximum level has been reached in the modern industrial societies to maximize economic benefit and social welfare (Benton, 2001). Therefore, there is a competition between the natural environment and the artificial environment to maximize economic benefit and social welfare in the modern industrial society (Tuna, 2006).

There are so many negative impacts on the natural environment in the modern industrial society. These impacts are mainly health problems and deaths as results of air, water, and soil pollution, global warming, global climate change, and climate polarization. These environmental problems give so many negative effects on the natural environment and the social environment. Therefore, these global environmental problems are not just accounted as physical environmental problems anymore; they should be accounted as societal problems. Most of these environmental problems are related to the social behavior of human beings in the modern industrial society (Tuna, 2006). The abovementioned problems might be evaluated as a reaction of the natural environment against its overusage, manipulation, and exploitation. Tourism activity is also one form of social behavior in the industrial society that affects the natural environment.

The degradation of the natural environment and the artificial environment are related to modernization and industrialization in the modern age. In other words, certain factors extend the reason of the mentioned degradation was modernized worldview. Moreover, exploitation and manipulation of the natural environmental resources has been legitimized by modernization ideology (Beck, 2000). As a result of that the society is alienated to the natural environment and especially the relationship between the natural environment and the society was weakened and almost broken down in the modern industrial society (Tuna, 2006). This situation is defined as alienation of the society to the natural environment. The environmental problems such as alienation of the society to the natural environment, degradation of the natural environment, environmental pollution, and related problems reached the maximum level in the second half the 20th century and these circumstances threatened the existence of human beings (Beck, 2001).

Therefore, some intellectual tendencies appeared to solve abovementioned environmental problems and to establish a peaceful relationship with the natural environment in the modern industrial society. This tendency called "back to nature" movement (Tuna, 2018). Ecotourism and reexploration of the natural environment by tourism are accounted as concrete examples of "back to nature" tendencies.

The tourism development as a multifaceted socioeconomic activity is mostly based upon a wider usage and manipulation of the natural environment. The relationship between the tourism development and the natural environment has two main dimensions. On the one hand, natural resources are main and the most critical resources to create a touristic destination; on the other hand, the tourism development gives so many negative impacts on the natural environment. Moreover, overuse and consumption of natural resources in the tourism processes have been changed the basic structure of the natural environment (Ceylan, 2001).

The tourism potential of a country is related to its social, historical, and cultural resources. Tourism is one of the most popular and economically most valuable sectors of the 21st century; however, the tourism development especially mass tourism is not so sensitive to the natural environment in the developing countries and the developed countries. As a result of disregard of the environmental impacts of tourism, the tourism development gives so many negative impacts on the natural ecosystem with threatening historical, structural, and natural resources of the area. Especially, faster tourism development and mass-tourism disqualified the natural environment. The faster tourism development gives so many negative impacts on the environment. These impacts are the buildings not compatible with the natural environment, destruction of coastal zones, esthetic pollution, insufficient infrastructure, sewage system and garbage collection, sea pollution, disposition of historical areas, degeneration of traditional social life. These impacts give greater impact on the natural environment such as changing land use, loosing agricultural lands and fauna and flora, and degradation of the natural environment (Öztürk, 2002).

The tourism development uses much more environmental resources than any other economic sector. The natural environmental resources are unavoidably important for the tourism development because they are not renewable, and they also take tourist interest.

The OECD report published in 1980, pointed out particularly important four environmental resources:

1. Wetlands: rivers, streams, seas, coastal zones, and other water resources.

2. Natural areas: soil, forests, and air.
3. Architectural areas: all architectural areas take tourist interests.
4. Social environment: all social activities.

Tourism is constructed on abovementioned resources. Moreover, an appearance and development of tourism in a certain area impacts abovementioned resources as well. Therefore, an environmentally friendly and naturally sensitive tourism development use the natural resources with more sustainable way and make tourism more sustainable (Mowforth and Munt, 1998). Therefore, the protection of natural resources as essence of the tourism development requires a high level of environmental sensitivity and environmental consciousness.

The tourism demand has been enormously increased since 1970s. There are both some positive and negative impacts of this demand. The development, management, and preservation of natural resources and cultural values of local communities have become major priorities for tourism development for the last few decades in Turkey. There is also a tendency to make these resources for marketable factors to the tourism development. Because tourists want escape from stressful working conditions and boring daily life of crowded metropolitan cities such as alienated lifestyle, overcrowded population, traffic and parking problems, and noise pollution. Then they want to spend their holiday in a peaceful atmosphere, nested in a natural environment. Therefore, a natural atmosphere is a basic demand for tourists and the tourism development (Mowforth and Munt, 1998).

Degradation of environment results with loosing competitiveness and resources related to the tourism development in Turkey. It is expected that a global demand for the environment is going to be increased in the near future. The environmental awareness of tourists is increasing and they more likely prefer environmentally friendly tourism products than others (CE, 1993).

9.9 THE SOCIAL AND ENVIRONMENTAL IMPACTS OF TOURISM DEVELOPMENT IN TURKEY

An enormous progress has been seen in terms of tourism development in Turkey for last three decades. While just a few hundred thousand foreign tourists had visited Turkey in the early 1970s, the number of foreign visitors reached more than 26 million in 2008 and more than 40 million expected in 2019. Additionally, the tourism sector generates more than 30 billion USD economical values and hundreds of thousands of jobs in Turkey. These basic figures present how fast and how big the tourism development in Turkey is. The positive economic impacts of tourism development are of course enormous. Turkish society enjoys this economic development and socioeconomic welfare generated by the tourism sector. However, there are so many unintended and unwanted social and environmental impacts of this development in Turkey which are not frequently investigated.

There is a wealthy literature on evaluation of the tourism development in Turkey (Tosun, 1999). Most of the studies look at the tourism development from an economical perspective and some of them look at an environmental perspective as well. However, the social impacts of tourism development in Turkey are not commonly investigated.

Initially, if it has been looked at from a sociological perspective, the social structure of the tourism cities shows interesting facts. The first and probably the most important point that should be mentioned about the social infrastructure of the tourism in Turkey is the education level of the tourism sector workers. The education level of the tourism sector workers is not so high (Tuna, 2007). Additionally, some of the tourism sector workers never had any occupational training course related to tourism. Therefore, some of the workforce has no idea about fundamentals of tourism such as tourist satisfaction, environment, and sustainability when they begin to work in the tourism sector. This low level of education is one of the most important problems of the tourism sector in Turkey. This circumstance has direct effect on the tourist satisfaction and sustainability of tourism development in Turkey. Therefore, an improvement of general level of education appears as a basic necessity for a more sustainable tourism development.

The second point that should be dealt with social structure of the tourism of Turkey is environmental sensitivity of the tourism sector workers. The environmental sensitivity and consciousness of tourism workers are so low in Turkey (Tuna, 2007). Not only workers but also managers of the tourism sector have not much knowledge about the environmental dimension and the environmental impacts of tourism development. This is also related to the general level of education. As general level of education of the tourism workers are so low, their environmental knowledge, sensitivity, and consciousness are low as well. Much more problems might be mentioned about the social structure of tourism in Turkey; however, abovementioned two problems crosscut all others. Therefore, only two of them are mentioned in this context.

It is possible to notice so many positive and negative impacts of tourism development in Turkey; however, most of the research points out positive impacts (Işıkçı, 2002). If it has been looked at impacts of tourism development in societal level; Turkish society most likely points out positive impacts of the development as well (Tuna, 2007).

Additionally, public perception on the tourism development shows contradictory results. Some sociocultural changes as a result of tourism development accounted with both their negative impacts and positive impacts simultaneously. For instance, dissolution of strong family ties and social solidarity both approved and disapproved by the same group of people (Tuna, 2007). On the one hand, Turkish society disapproves the dissolution of strong family ties and social solidarity because of acceptance of these values as the distinctive features of the traditional societies and Turkish society is still somewhat traditional. On the other hand, Turkish society approves the dissolution of strong family ties and social solidarity. Because these changes accepted as indicators of individual freedom, modern family life, and the modern lifestyle (Giddens, 1991). Turkish society accepted these values as an indicator of individual freedom as a modern society (Gouldner, 1962). This circumstance might be acknowledged as an indicator of a dual social structure or a social change from a traditional social structure to a modern one (Harrison, 1995). Moreover, tourism is a main catalyst or agent of the social change or the dissolution process (Dann and Cohen, 1996).

However, the dissolution of strong family ties and social solidarity has not been easily accepted as social changes by the individuals of Turkish society. On the one hand, strong family ties and compassions about social solidarity are being dissolving as major social institutions of the traditional society. On the other hand, counterinstitutions, such as community clubs, youth clubs, family clubs, and veteran clubs and advisory boards, and value systems of these institutions have not been established and institutionalized synchronically as major institutions of modern society. Therefore, individuals in the society feel like normlessness, valuelessness, and even alienation. The social control mechanisms have become ineffective. These sorts of problems have been experienced in Turkish society related to a swift increase in income as a result of tourism development. The relationships between the family members, especially the parents and the youngsters are being dissolved. There is a tendency through alcoholism, drug usage, and underage sexual intercourse in Turkish society that are not legitimately acceptable behaviors in a traditional society. All these social changes might be acknowledged as a sociocultural cost of a swift tourism development in a traditional society (Dann and Cohen, 1996).

Therefore, the problems of the erosion of traditional social values with the tourism development and lack of compensation of these values with new modern social values not always explicitly have been seen. However, these problems culminated with passionately dissatisfactory feelings in Turkish society. The community consultation centers should be established to resolve abovementioned problems. The local governments, occupational associations, and nongovernmental organizations may play initiative roles to establish these organizations. Additionally, universities play an advisory role to institutionalize these consultancy services.

Most of the tourism areas have been established on the productive agricultural areas in Turkey. The coastal tourism areas of Turkey are located on the agriculturally productive the Mediterranean and the Aegean regions. So many agricultural lands had been eaten by the tourism development to build up new tourism facilities. Therefore, productive agricultural lands have been switched into tourism areas in these regions. As a result, the agricultural potential of Turkey also decreased related to the tourism development (Dindar, 2002).

The tourism development also affected the urban and rural structure, especially in the coastal areas of Turkey. Small agricultural and fisheries towns have become large tourism centers for the last 20–30 years especially in the coastal areas. The wetlands, foresters, lakes, rivers, and seas have been deteriorated. The building construction enormously increased, many touristic facilities and houses have been constructed in considerably small areas and the carrying capacity has been passed over for this period. This overintensive construction results with many infrastructural problems, such as insufficient roads, water systems, and sewage systems. The construction of these infrastructures is needed huge amount of budgetary and financial resources. Because of the lack of financial resources these basic infrastructures are not sufficient. Especially, insufficient sewage systems result with water pollution in the sea and other water resources. The overintensive urbanization in the touristic city centers also created parking problems, air pollution, esthetic pollution, and noise pollution (Tuna, 2007).

Therefore, to change tourism structure to a more sustainable way, a strict control over new urbanization tendencies is a fundamental necessity in the tourism areas. No more building permit will be given to control over intensive construction in the city centers. New tourism development and rehabilitation plans should be designed if it is necessary in the tourism areas. Some of the tourism areas already announced as specially protected areas in Turkey. However, because of the inefficient control mechanisms of related institutions and lack of cooperation among the public and private stakeholders such as local governments, local branches of central government, NGOs, and tourism organizations, mentioned protection plans have not been applied effectively. Therefore, initially an effective cooperation among the mentioned institutions should be institutionalized. Then, these areas should strictly be protected for a sustainable tourism development in Turkey. General, regional, and local environmental protection plans should be designed and applied for a sustainable tourism development in Turkey. The objectives of regional and local environmental protection plans are as follows:

1) prevention of sea pollution,
2) prevention of algae and dead seeds in the sea,
3) crime control and prevention,
4) prevention of high population density and overcrowded population,
5) prevention of noise pollution,
6) distinction of tourism areas and nontourism areas,
7) distinction of rural areas,
8) prevention of overconstruction and overurbanization in the tourism areas, and
9) a restoration of negatively urbanized areas.

Local-level and regional-level environmental management systems should be institutionalized to realize abovementioned objectives.

These are the principle of environmental management systems:

1) formulating of environmental policy,
2) ensuring total commitment of all in the organization,
3) carrying out environmental review,
4) determination of responsibilities within the organization,
5) preparing a register of environmental effects,
6) establishing objectives and targets,
7) implementing management systems,

8) commissioning periodic environmental audits,
9) performing regular systems reviews based on performance (Kirk, 1995).

9.10 CONCLUSION

Tourism is one of the most important economic sectors in Turkey. There is a faster development witnessed in the tourism sector in Turkey for the last few decades. Tens of millions of foreign tourists are visiting Turkey; hundreds of thousands workforce are working on the tourism sector and tens of billions of USD revenues are generating in the sector. Therefore, there is an enormous economic impact of tourism development in Turkey. Turkish society enjoyed economic value that is generated by the sector. This development increased Turkish society's social and economic welfare. Accordingly, general views of the coastal towns of the Aegean and the Mediterranean regions have been changed by the tourism development. Many infrastructural investments such as roads, water systems, and sewage systems have been realized for the last few decades in the mentioned regions (Tuna, 2011).

However, there are many unwanted and unintended social and environmental impacts of the development. The basic features of traditional social structure such as strong family ties and strong social solidarity have been dissolved. On the contrary, new modern social institutions such as family, youth, and retirement consultation centers have not been established. The traditional social institutions evaporated; however, modern social institutions have not been established. As a result of these circumstances, the Turkish society feels like normlessness and valuelessness.

There is an enormous change have been seen in the natural environmental scenery in Turkey for the last few decades. Water pollution, sea pollution, land pollution, degradation of the natural environment, over intensive population, loosing countryside and agricultural areas are major environmental problems that have been witnessed in Turkey. Therefore, to take necessary measurements and to initiate appropriate policies are major requirements to prevent the negative social and environmental impacts of the economic development in Turkey. The policies of social and economic development should be accorded with sustainability principles.

Particularly, an establishment of general, regional, and local level environmental management plans appears as necessity for a sustainable tourism development in Turkey.

KEYWORDS

- **Turkey**
- **tourism development**
- **environment**
- **environmental impacts**

REFERENCES

Apostolopoulos, Y. Introduction: Reinventing the Sociology of Tourism. In *The Sociology of Tourism: Theoretical and Empirical Investigations*; Apostolopoulos, Y., Leivadi, S., Yiannakis, A., Ed.; Routledge: London and New York, 1996.

Baudrillard, J. *The Consumer Society: Myths & Structures*; Sage Publications: London, 1998.

Beck, U. Risk Society Revisited: Theory, Politics, and Research Programmes. In *The Risk Society and Beyond: Critical Issues for Social Theory*; Adam, B., Beck, U., Loon, L. V., Ed.; Sage Publications: London, 2000.

Benton, T. *Environmental Sociology: Controversy and Continuity*; Sosiologisk Tidsskrift, 2001.

Çağatay, A.; Yurdaer, M.; Kırış, R. Ekoturizm için mekansal ve yerel toplulukların katılımının planlanması (Akseki-İbradı havzası örneği) *III. Turizm Şurası Bildirileri, Turizm Bakanlığı Yayını,* Ankara, 2002.

CE (Commission of the European Communities), Directorate General XXIII-Tourism Unit. *Taking Account of Environment in Tourism Development*; Commission of the European Communities: Brussels, 1993.

Ceylan, T. *Turizm ve sürdürülebilir gelişme*; *Anatolia*. Yıl: 12, Ankara, 2001.

Cohen, E. A Phenomenology of Tourist Experiences. In *The Sociology of Tourism: Theoretical and Empirical Investigations*; Apostolopoulos, Y., Leivadi, S., Yiannakis, A., Ed.; Routledge: London and New York, 1996.

Dann, G.;Cohen, E. Sociology and Tourism. In *The Sociology of Tourism: Theoretical and Empirical Investigations*; Apostolopoulos, Y., Leivadi, S., Yiannakis, A., Ed.; Routledge: London and New York, 1996.

Dindar, M. *Turizm ve çevre ilişkileri III. Turizm şurası bildirileri*. Turizm Bakanlığı Yayını: Ankara, 2002.

Eckersley, R. *Environmentalism and Political Theory*; State University of New York Press: New York, 1992.

Giddens, A. *Modernity and Self-Identity in the Late Modern Age*; Cambridge: Polity Press: UK, 1991.

Giddens, A. *The Consequences of Modernity*; Cambridge: Polity Press, 1990.

Gouldner, A. Anti-Minotaur: The Myth of a Value Free Sociology. *Soc. Probl.* **1962,** 9, 199–213.

Göymen, K. Tourism and Governance in Turkey. *Ann. Tourism Res.* **2000,** *27* (4), 1025–1048.

Harrison, D. *Sociology of Modernization and Development,* Routledge: London, 1995.

Harper, C. L. *Environment and Society: Human Perspectives on Environmental Issues*; Parentice Hill: New Jersey, 1996.

Hildebrand, P. M. The European Community's Environmental Policy, 1957 to 1992: From Incidental Measures to an International Regime? *Environ. Polit.* **1992,** *1* (1), 13–43.

Inforegio/Panorama. *Sustainability—Holy Grail of Tourism in the 21 Century, European Union Regional Policy, Cooperation without Borders*; Office for Official Publications of the European Communities: Luxembourg, January 2002/2006.

Işıkçı, Y. *Türk turizminde altyapı ve çevre iii. Turizm şurası bildirileri*; Turizm Bakanlığı Yayını: Ankara; 2002.

Judge, D. A Green Dimension of the European Community. *Environ. Polit.* **1992,** *1* (1), 1–9.

Karaman, M. R. Eko-turizm yaklaşımında verimli toprakların korunması ve ekolojik dengenin önemi, *III. Turizm Şurası Bildirileri, Turizm Bakanlığı Yayını*; Ankara, 2002.

Karamustafa, K. Marketing-Channel Relationships: Turkey's Resort Purveyors' Interactions with International Tour Operators. *Cornell Hotel Restaur. Adm. Q.* **2000,** *41* (2), 21–31.

Kasapoğlu, M.; ve Ecevit, M. Attitudes and Behaviour toward the Environment: The Case of Lake Burdur. *Environ. Behav.* **2002,** *34* (3), 363–367.

Kirk, D. Environmental Management in Hotels. *Int. J. Contemp. Hospitality Manage.* **1995,** *7* (6), 3–8.

Korzay, M. In *Turkish Tourism Development, Tourism: The State of the Art*; Seaton ve diğerleri, A. V.; Wiley: Chichester, 1994; pp 85–99.

Lanquar, R. *Turizm ve seyahat sosyolojisi*. İletişim Yayınları, İstanbul; 1999.

Lenski, G. E. *Power Privilege: A Theory of Social Stratification*; McGraw-Hill Book Company: New York, 1966.

Milbrath, L. W. *Environmentalists: Vanguard for New Society*; State University of New York Press: Albany, 1984.

Mowforth, M.; Munt, I. *Tourism and Sustainability: New Tourism in the Third World.* Routledge: New York, London, 1998.

Naess, A. Deep ecology. In *The Green Reader: Essays toward a Sustainable Society*; Dobson, A., Ed.; Mercury House, Incorporated: San Francisco, 1991.

Özdemir, Ş. *Türkiye'de toplumsal değişme ve çevre sorunlarına duyarlılık*; Palme Yayıncılık: Ankara, 1988.

Özdemir, M. *Turizmde fiske taşları*; Rekmay, Ankara, 1998.

Özgür, Ö. *Bitmeyen Kavga Gökova*; Özlem Matbaası: Muğla, 1998.

Öztürk, S. Eko Turizm Destinasyonu Olarak İznik, *III. Turizm Şurası Bildirileri, Turizm Bakanlığı Yayını,* Ankara, 2002.

Swarbrooke, J.; Horner, S. *Consumer Behaviour in Tourism*; Butter-Worth Heinnemam: Oxford, 2001.

Tosun, C. An Analysis of the Economic Contribution of Inbound International Tourism in Turkey. *Tourism Econ.* **1999,** *5* (3), 217–250.

Tourism and the Environment in Europe, *An Introduction along with Reports on Some European Projects in Tourism*, Tourism Division of the European Commission, DG XXIII Alpensfors-chunginstitut gemn., GmbH Garnish-Partenkirshen, Germany, 1995.

Tuna, M. Environmentalism: An Empirical Test of Multi-Effects on Environmental Attitudes in More and Less Developed Countries, Unpublished Dissertation, Mississippi State University, 1998.

Tuna, M. *Yatağan termik santralinin çevresel ve toplumsal etkileri*; Muğla Üniversitesi Yayını, Muğla, 2001.

Tuna, M. Globalization of Environmentalism: World Environmentalism System. *Paper presented at the XV World Congress of Sociology, Brisbane, Australia*, 2002. http//203.94.129.73/docs/p1702.rtf.

Tuna, M. Public Environmental Attitudes in Turkey. In *Environmental Justice and Global Citizenship*; Phillips, C., Ed.; Interdisciplinary Press: Oxford, United Kingdom, 2004.

Tuna, M. Cultural Approximation and Tourist Satisfaction. In *Progress in Tourism Marketing*; Kozak, M., Andreu, L., Ed.; Elsevier: London, 2006a.

Tuna, M. *Türkiye'de çevrecilik: Türkiye'de çevreye ilişkin toplumsal eğilimler*; Nobel Yayın: Ankara, 2006b.

Tuna, M. *Turizm, çevre ve toplum: Marmaris örneği*; Detay Yayıncılık, Ankara, 2007.

Tuna, M. *Globalization of Environmentalism: Impacts of Multi-Level Effects on Environmental Attitudes*; Lambert Academic Publishing: Saarbrücken, 2010.

Tuna, M. Social and Environmental Impacts of Tourism Development in Turkey. In *Sustainability of Tourism: Cultural and Environmental Perspective*; Kozak, M., Kozak, N., Ed.; Cambridge Scholars Publishing: UK, 2011.

Tuna, M. Society, Environment and Tourism Relationship in Turkey. *Turkish Social Structure,* Anadolu Üniversitesi Yayınları: Eskişehir, 2018.

Wallerstein, I. *The Capitalist World Economy*; Cambridge University Press: Cambridge, 1979.

Weale, A.; Williams, A. Between Economy and Ecology? The Single Market and the Integration of Environmental Policy. *Environ. Polit.* **1992,** *2* (1), 45–63.

INDEX

U

W